Quantitative Methods for Studying Design Protocols

Jeff W.T. Kan · John S. Gero

Quantitative Methods for Studying Design Protocols

Springer

Jeff W.T. Kan
Department of Architecture and Civil Engineering
City University of Hong Kong
Hong Kong

John S. Gero
Computer Science and Architecture
University of North Carolina at Charlotte
Charlotte, NC
USA

and

Krasnow Institute for Advanced Study
George Mason University
Fairfax, VA
USA

ISBN 978-94-024-1456-1 ISBN 978-94-024-0984-0 (eBook)
DOI 10.1007/978-94-024-0984-0

Softcover reprint of the hardcover 1st edition 2017

Printed on acid-free paper

This Springer imprint is published by Springer Nature
The registered company is Springer Science+Business Media B.V.
The registered company address is: Van Godewijckstraat 30, 3311 GX Dordrecht, The Netherlands

Preface

This book addresses the development of quantitative tools intended to improve the understanding of design activities. The motivation of the research was to understand human design processes and test new approaches to analyse design protocols.

Two foundational concepts are presented: (1) an ontologically based coding scheme and (2) linkography that utilises that coding scheme. The Function-Behaviour-Structure (FBS) ontology is used as a foundation to understand designing and to examine design protocols—protocols are segmented and coded according to FBS design issues. This provides a coding method that is independent of the design domain, the number of designers, whether the designers are collocated or not, and whether or not they use tools. Linkographs are composed of the semantic connections between segments. Since each segment is already coded with an FBS design issue, a link defines the transformation of one FBS design issue to another—in other words—a design process. Such a linkograph generates all the design processes in a design protocol. The coded protocol and its linkograph provide the base datasets for a range of analysis techniques including standard statistics, Markov modelling and clustering. Entropy, as it is used in information theory, is used to measure linkographs and test for correlations between entropy and design outcomes.

These quantitative tools were developed and tested on data from a pilot study and from various design experiments. The design protocols presented in this book are for demonstration purposes and the results from them are indicative of the kinds of results possible rather than generalizable conclusions. The concepts and techniques described in this book to examine design activities are aimed at increasing the tools and techniques available to readers to increase their understanding of this activity called designing.

Hong Kong, Hong Kong — Jeff W.T. Kan
Charlotte, USA — John S. Gero

Contents

Chapter 1
Introduction

Designing is one of the most profound of cognitive activities. Humans intentionally change their surrounding—physical, social and virtual worlds—through design, so it is surprising that our understanding of it is limited compared to other disciplines. The concept of design is multifaceted; in this book, we consider it as an intentional act, rather than a unitary activity, that consists of definable processes. This chapter introduces the motivation behind, and the structure of, this book.

1.1 Why a Book About Quantitative Methods and Design Protocols?

The overarching motivation of this book is to assist in the understanding of human design processes through protocol analysis. There is currently a lack of unified knowledge about this unique human capability—designing. The book revolves around two closely related questions. Can good design be identified during the design process rather than merely assessing the aftereffects? And if so, what kind of apparatus can we use to measure design processes? In classical protocol studies of designers, many coding schemes have been developed that are unique to the data. This approach limits the applicability of the results obtained. Is there a more general coding scheme that can be reused in widely varying circumstances?

1.1.1 Understanding Designing

Designing is an important activity that is often used to distinguish humans from other beings. However, our knowledge of the cognitive processes of designing

J.W.T. Kan and J.S. Gero, *Quantitative Methods for Studying Design Protocols*, DOI 10.1007/978-94-024-0984-0_1

remains sparse. In recent years, design thinking and research have been popular in many fields (Lockwood 2009; Plattner 2012; Ingle 2013; Cross 2011). However, there is no agreement about the methodologies for studying designers. Protocol analysis has become one of the de facto method for studying the cognitive processes of designers for many years. Cross et al. (1996a, p. 1) stated:

> Of all the empirical, observational research methods for the analysis of design activity, protocol analysis is one that has received the most use and attention in recent years. It has become regarded as the most likely method (perhaps the only method) to bring out into the open the somewhat mysterious cognitive abilities of designers.

There are other research methods, which can be considered as subsets of protocol studies of designers that use verbal data as the raw material for analysis such as conversation analysis (Oak 2011) and computation linguistic analysis (Dong 2009).

Protocol analysis has been used qualitatively (Cross and Cross 1996; Lloyd 2007) and quantitatively (Eastman 1970; McNeill et al. 1998; Gericke et al. 2007). It has also been used in both in situ (Zolin et al. 2004; Olson and Olson 2000) and in vitro conditions (Gabriel 2000; Kan and Gero 2008a). Dorst and Dijkhuis (1995) suggested that there are two basic types of analysis, based on two different paradigms—namely, content-oriented and process-oriented analysis. The process-oriented analysis is rooted in viewing designing as a form of information processing (Eastman 1970). The content-oriented analysis derives from a perception of designing as a conversation with the materials (Schon and Wiggins 1992). The classical quantitative methods of analysis, based on the foundational text by Ericsson and Simon (1993), mainly rely on tabular statistics and the correlation of coded segments. The coding scheme very much depends on the focus of the study and is usually developed in an ad hoc manner, based on the data. As a result, it is difficult to compare or measure results drawn from different research studies, hence the advancement of the understanding of designing remains not well-founded. The Design Thinking Research Symposia 2 (DTRS2) (Cross et al. 1996b) and 7 (DTRS7) (McDonnell and Lloyd 2007) are examples of how different researchers, each with their own particular view of designing, employ different methods and coding schemes to study the same protocol data. As the content of the data between DTRS2 and DTRS7 are quite different, the analysis methods are more diverse in DTRS7. The methods used in DTRS7 range from narrative inquiry and conversational analysis, to linguistic analysis. This diversity facilitates a rich and broad understanding of different aspects of designing. However, it also makes it difficult to compare the results from different research and obtain a deeper understanding of designing.

1.1.2 Design Assessment

The assessment of design objects is essential for improvement. However, the often-subjective nature of design assessment presents a challenge to develop objective measurements. Design is usually assessed in terms of its artefacts but there is an interest, especially in design education, to have some kind of predictive test that is based on the design process. For example, Goldschmidt (1992, 2014) attempted to find the relationship between design performance and the integration of ideas; van der Lugt (2003) correlated the goodness of idea with the types of interconnections among ideas during idea-generation meetings; Kvan and Yunyan (2004) tried to correlate learning styles, which influence the design process, and studio grades; and Goldschmidt and Tatsa (2005) found an indicative correlation between influential ideas and students' project grades. Kruger and Cross (2006) studied the effect of design strategies upon design outcome. These studies not only provide conjectures about the inquiry methods of assessing designing but also indicate some conditions that contribute to good designing. However, empirically verifiable research concerning the relationship between the design process and design outcomes that builds on previous research, remains scarce.

1.1.3 Tools for Studying Designing

As there is an increased interest in protocol analysis in different fields (Cross et al. 1996a), software support has recently been developed to code and analyse protocol data. However, this tends to follow a certain view about studying protocol data and, when used to study designing, it mainly supports the process-oriented type of analysis. For example, "Mangold Interact", a popular protocol analysis software, uses a linear time frame to segment protocol and allows codes to overlap. The use of this paradigm will provide insight into certain aspects of designing; however, this method may not correspond to other aspects of designing, such as the nested nature of idea-generation activities. Permitting overlapping of codes, with the same or different segment lengths, encourages the spontaneous addition of codes, which helps to capture the current state of affairs, but introduces incommensurability. Are there other tools to study design protocols? Will a unified coding scheme be applicable to the study of designers in different situations? Can the design outcome be discerned from the design process?

1.2 Structure of the Book

In this book we develop and test new approaches to the quantitative analysis of design protocols that will increase our understanding of designing. New techniques are developed based on information theory, sequential analysis and design ontology, which are then verified with empirical data. These new techniques open up the possibility of investigating protocol data in a quantitative manner (measures of diversity, time-related events and derived design processes) augmenting the possibilities offered by the treatment of protocol data using standard statistics (correlation and other measures and tests).

Chapter 2 reviews the related literature and defines the scope of this book.

Chapter 3 introduces the four tools used for this investigation, namely: linkography, information theory, design ontology and Markov analysis. The construction of a linkograph, a graphical representation of the linkages among the moves or segments of a design protocol, is examined. This chapter describes how a linkograph is used to capture designing information, which is then analysed using statistics and clustering. The foundation of using entropy, as in information theory, to measure linkographs is explored. The hypothesis is that some entropy correlates with the design outcome. An ontologically derived coding scheme—the Function-Behaviour-Structure (FBS) scheme is described in this chapter to enrich the results by adding semantics to the linkograph. The use of Markov analysis as a tool to examine sequential events is also presented.

Chapter 4 reports on two case studies in which the FBS ontology and linkography are applied to study different types of design sessions. It involves the classification of segments in terms of FBS classes and the derivation of processes from the linkograph.

Chapter 5 presents a study of two collaborative design sessions in two different situations, one a face-to-face (in situ) session and the other a computer-mediated session that simulates distance collaboration (in vitro). The protocols are analysed by the methods proposed in Chapter Three. Results from these analyses are presented in this chapter.

Chapter 6 provides an account of an experiment with twelve design sessions. Linkographs are constructed and entropic measures are taken to determine whether they correlate with the design outcomes.

Chapter 7 concludes the findings of this study and discusses the usefulness of individual techniques.

Chapter 2
Background

This chapter provides a critical analysis of the relevant literature concerning the study of designers, protocol studies of designers and the measurement of designing. It defines the scope of the material in this book and clarifies terminologies such as creativity, design, designing, design collaboration and design assessment which are used in subsequent chapters. Various models and theories about design have been developed, and different claims have been made, based on case studies, concerning the creativity and quality of design. Instead of reviewing these models and claims, the focus of this chapter concerns methods of inquiry.

2.1 Studying Designers

Designing is one of the most significant intentional acts of human beings and is viewed as one of the most complex of human endeavours. Positions and philosophies of "what is design" determine how it is studied. Many designers develop their own philosophy and style. Various design strategies are available, such as using a first principle to find design solutions, analogies and metaphors, previous cases, delaying decisions until a design emerges, interacting to stimulate ideas, and so forth. Herbert Simon's classic work on artificial intelligence (Simon 1969) had a strong influence on design research; "designing as problem solving dominated the thinking of design researchers" (Gero 2007, p. 17). Others researchers (e.g., Cross 2007a; Jones 1970) viewed designing as a cross disciplinary field that embraces the humanities, the sciences, mathematics and art.

This book argues that the study of designers can be empirically based and yields an understanding of the cognition of designers. The methods to study the cognition of designers fall into five methodological categories: questionnaires and interviews (Cross and Cross 1998; Lawson 1997; Murty 2006); input-output experiments (where the designer is treated as a black box which produces the behaviors in the

J.W.T. Kan and J.S. Gero, *Quantitative Methods for Studying Design Protocols*, DOI 10.1007/978-94-024-0984-0_2

outputs for changes in inputs) (Purcell et al. 1993), anthropological studies (Lopez-Mesa and Thompson 2006), introspection (Galle and Kovacs 1992) and protocol studies. Protocol studies can be further divided into retrospective protocol studies (Suwa et al. 2000), and concurrent protocol studies (Eastman 1970). Protocol analysis is currently the useful method of studying designers (Cross et al. 1996a).

2.1.1 Protocol Analysis

Protocol data, based on samples of observations, are essentially qualitative. Ericsson and Simon (1993) laid the foundation of using verbal protocols, concurrent reporting, as quantitative data to study thought processes; van Someren et al. (1994) provided the theoretical background and a practical guide for the study and modelling of cognitive processes. They assumed a simple human cognitive model, Fig. 2.1, to develop the validly of verbal reports. The arrows in the diagram represent five different cognitive processes: perception (from sensory to working memory), retrieval (from long-term memory to working memory), construction (within working memory), storage (from working memory to long-term memory), and verbalisation (from working memory to protocols). The sources of invalid and incomplete verbal data were identified as being disturbance of the thought process, memory errors and the interpretation of the cognitive process. Their study also provided some practical guidance on how to obtain "good" protocols during experiments, such as taking care of subjects' feelings by offering assurances that their privacy would be protected; giving clear instructions to the participants; keeping the participants "thinking aloud", without disturbing or forcing them to interpret; and arranging a warm-up session involving similar tasks, to encouraged the participants to practice their verbalising concurrently.

The disadvantages of concurrent reporting are: (1) it slows down the thought process; (2) it does not review the whole thought process when the participants stop verbalising or use imagery only; (3) it impairs reasoning for those participants who cannot verbalise and reason simultaneously; and (4) there are subjective elements in the coding of concurrent reporting.

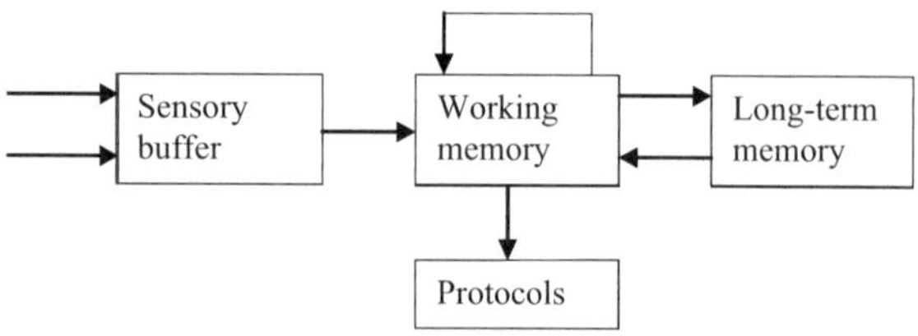

Fig. 2.1 van Someren's memory model (van Someren et al. 1994)

The fourth drawback can be improved by techniques such as inter-coder arbitration. Retrospective reporting with visual aids can prevent any slowing down of participants' reasoning while verbalising. These visual aids include the artefacts that the participant produced and the video recording of the participant designing.

The study of design thinking has been characterised as a method somewhere between the hard sciences and the social sciences (Cross 2007b). Protocol analysis has been used to identify different design activity, reveal different mental models and the knowledge structures of designers, as well as to investigate the perceptual aspects of sketching and designing (Atman et al. 2007; Tang 2002; Yilmaz et al. 2015). According to Akin (1998) the first formal protocol analysis of designing was conducted by Eastman (1970). Eastman's study contributed to the current understanding of what designers do when they design in the form of an information process model. Eastman viewed designing as a process whereby problems are identified and alternative solutions are tested. This view was challenged by a view of designing as a reflective conversation with material (Schon and Wiggins 1992) in which the basic structure is an interaction between designing and discovering.

Protocol Analysis in Different Design Domain

Besides the architecture design domain that Charles Eastman and Donald Schon applied protocol analysis to investigate design activities, researchers in other domains also use this technique to examine designing. In engineering design education, Atman and Bursic (1998) used protocol analysis as a tool to assess student's design process so as to evaluate the impact of changes in engineering education program so as to improve the way open-end design is being taught.

In engineering systems design, Ennis and Gyeszly (1991) studied six experienced designers solving engineering packaging problems. Verbal protocol analysis was used to identify how the designers introduced information or knowledge into the design process. They found that gathering information was a crucial approach for these expert designers to solve design problems and generate design ideas.

Hughes and Parkes (2003) surveyed the use of protocol analysis in software engineering research from the 1980s to 2003. Their conclusion was that the protocol analysis technique "has contributed towards the development and testing of models of the information processing during the software engineering process, particularly those relating to software design and comprehension" (Hughes and Parkes 2003, p. 138). However they also found difficulties associated with this method, they included: "(1) the effort of devising a valid and reliable encoding scheme; (2) the time-consuming nature of the encoding process; and (3) the problem of comparing results from researchers who have applied different encoding schemes" (Hughes and Parkes 2003, p. 138).

Process-and Content-Oriented Protocol Analysis

Dorst and Dijkhuis (1995) suggested that there are two types of analysis in protocol studies, namely, process- and content-oriented protocol analysis. Each captures the two different paradigms mentioned earlier—the information processing model and the reflection in action model. Usually the think-aloud or concurrent protocol is used for process-oriented analysis, in which the processing of information is the focus. The retrospective protocol is often used for content-oriented analysis, in which the focus is on the content of designing. However, increasingly both protocol methods are used for both purposes.

Tang (2002) carried out a detailed empirical comparison between the retrospective and the concurrent protocols. He found they were similar in terms of quantity and quality; quantity related to the number of segments. In terms of quality, the concurrent protocols revealed more information related to the functional aspect of designing, whereas the retrospective protocols revealed more information about producing solutions and evaluation. There are many differences in the approaches to methods and the coding schemes coupled with specific aspects of both content and process-oriented protocol studies. The next subsection reviews a number of issues related to the commonly used protocol analysis of designing.

2.1.2 Measurement of Design Protocols

The analysis of design protocols, both content- and process-oriented, or any type of design protocol analysis usually involves statistical methods.

Unit of Analysis

The unit of analysis varies according to the objectives and foci of studies. It can be individual participants (in the study of design teams), sessions, episodes, code categories, or even each segment/utterance. In the classical method of studying design protocol, van Someren et al. (1994) classified the procedures into five steps: conducting experiments, transcribing protocols, parsing segments, encoding according to a coding scheme, and interpreting the encoded protocols. The first step is derived from the research aim and method.

Diversity in Segmenting

In parsing segments, there are different ways to segment protocols, depending on the objectives and scope of the study. For instance, protocols can be segmented according to instances of processes in order to study the frequencies of processes. Ericsson and Simon (1993) suggested that appropriate cues for segmentation are pauses, intonation, and contours, which correspond to their information processing model. Gunther et al. (1996) along with Dorst and Dijkhuis (1996), used a fixed 15-second time-scale. The advantage of this method is that it requires no interpretation; hence it quickly segments the protocols. However, the obvious problem with a fixed time-scale is that it may cut in the middle of a statement, which could make the coding difficult; therefore additional criteria are required to handle these cases.

Another way of segmenting protocols relates to the designers' lines of intentions or actions (Gero and McNeill 1998; Suwa et al. 1998). In this category, there are also differences in whether the categorisation affects the segmentation. In Gero and

McNeill (1998), one sub-category corresponds to one segment. On the other hand, Suwa et al. (1998) proposed that one segment might contain several sub-categories.

Yet another way to segment design protocols is by "design moves", which Goldschmidt (1990) introduced as "the smallest coherent operation detectable in design activity" (Goldschmidt 1992) but Perry and Krippendorff (2013) in their study found that identifying the boundaries of the move was not reliable with student coders even with training.

Diversity in Coding Schemes

Code categories are defined by a coding scheme, many of which have been developed for use with design protocols. All such schemes are based on particular views of the activity of designing, and most are unique to the data to which they are applied. For example, to document engineering student design process, the coding scheme (Atman and Bursic 1998) used contains four main variables: design step, information processed, activity and object. Within each variable there are sub-categories of codes, for example in the design step there codes for need, problem definition, gather information, generate ideas, modelling, feasibility analysis, evaluation, decision, communication and implementation.

In Hughes and Parkes' (2003) survey of protocol analysis in software engineering research, they grouped specific sub-domain or activities related to: requirement analysis, design meetings, debugging, re-engineering, corrective maintenance and team reviews; within which the study areas were focused in team work, novice vs expert, debugging strategies, domain knowledge, etc. They found that early work focused efforts to devise a cognitive model of programming behaviour, and gave some attention to different strategies used with different programing languages. Attention moved then to examine the software design process and use of tools to support the designing. Later research studied alternative methodologies and modelling design processes. They also found two recurring themes: (1) an investigation of the design process and (2) a comparison of behaviours between levels of expertise. They noticed that though specific themes have recurred throughout the two decades that they surveyed, no common coding scheme has been developed that can be applied in a range of different circumstances. They also evaluated a general-purpose coding scheme, 'A Flexible Expandable Coding Scheme' (AFECS) (von Mayrhauser and Lang 1999), and found it is helpful as a template to determine the basic structure of a coding scheme but the actual coding scheme constitute a fraction of a customised general-purpose coding scheme.

Gero and McNeill's (1998) developed one of the most comprehensive process-oriented coding schemes concerning designing, which contains multi-dimensional categories. One dimension of the categories concerns the designer's navigation within the problem domain with different levels of abstractions. Another dimension concerns the strategies used by the designer. Yet another dimension relates to the designer's reasoning about function, behaviour or structure.

Suwa et al.'s (1998) coding scheme is a good example of a content- oriented coding scheme. It was based on the human cognitive process sequence—sensorily, perceptually, and then the semantic categorisation of design actions into four categories: (1) physical, corresponding to the sensory level, consists of categories of making depictions, examining previous depictions and other physical actions; (2) perceptual, corresponding to perceptual action, contains categories of attending to visual features, attending to spatial relations, and organising or comparing; (3) functional, contains categories of design artefacts: issues of interaction, and psychological reactions of people; and (4) conceptual, corresponding to the semantic, consists of categories such as making evaluations, establishing goals and retrieving knowledge. This coding scheme was not originally designed to study group or team designing.

All the above research suggested that it is difficult to have a general and efficient coding scheme to map different design situations and scenarios onto design processes because of the diversity of domains and various views of designing.

Design Teams

There is an increased interest in understanding team designing processes and activities. It is impractical to undertake retrospective studies when a design team consists of more than three members. Compared to individual designing, studies (Cross and Cross 1995; Gabriel 2000; Olson and Olson 2000; Zolin et al. 2004) have shown that there are a multiplicity of factors that contribute to or affect team designing. Some of these factors are role and relationship, trust, social skills, common ground, organisational context and socio-technical conditions. Most of these factors are underpinned by communication, either verbal or non-verbal. Cross et al. (1996a) suggested that the verbal communication of members provides indicative data on their cognitive activities.

The protocol analysis technique has been adopted to understand the interactions of design teams (Cross and Cross 1996; Mazijoglou et al. 1996; Stempfle and Badke-Schaub 2002) and the behaviour of teams (Goldschmidt 1996; Valkenburg and Dorst 1998). Bly and Minneman (1990), along with other protocol studies (Gabriel 2000; van der Lugt 2003) suggested that with the introduction of technology, designers will adapt their activities accordingly. Goldschmidt (1996) and van der Lugt (2003) both used linkography (explained in the next section, Sect. 2.1.3) as a base for their studies. Valkenburg and Dorst (1998) used a similar, albeit differently presented, method to trace reflection in action by relating (rather than linking) segments in a protocol in terms of naming, framing, moving and reflecting.

Although some researchers use similar methods, there is no unified framework that can be applied to the study of design teams. The existing protocol analysis methods developed to study designers may need to be revisited if we are to understand the dynamics of team designing and then compare them to individual designing.

Statistical Analysis of Design Protocol

In statistical terms, the coded segments of protocol data usually contain two parts: the qualitative part with categorical (nominal) data and the quantitative part concerning duration (time). There are generally two types of analysis—descriptive statistics and inferential statistics—both based on the assumption of distribution.

Descriptive statistics are used to summarise the protocol data—usually in the form of charts and tables. This kind of analysis can reveal how the designers spent their time. For example, Maher et al. (2006) used descriptive statistics to study the impact of a collaborative virtual environment on design behaviour. They found that the designers spent the largest percentage of their time focused on communicating about the design task and on actions to produce an external representation in all environments.

Inferential statistics are used to test models of designing from protocol data. For example, hypothesis-testing can verify proposed models of designing. McNeill et al. (1998) used a *t* test on the hypothesis that the design process moves from a design requirement, expressed in terms of function, to a design description couched in structural terms. They also used linear regression to test the transition relationship between "evaluation to analysis". Hypothesis-testing was also used to compare designers working in different conditions, or to compare different designers working in similar conditions. The chi-square test is another common tool used in protocol analysis for hypothesis testing. It tests if the frequency distribution of certain coding categories observed in a protocol is consistent with a particular theoretical distribution. Readers can refer to standard text on statistics for the concepts and conditions behind these tests.

Relationships among variables and categories can also be explored with a statistical method. For example, Kvan and Yunyan (2004) correlated students' learning styles with their performance in the design studio. Kavakli and Gero (2002) used a correlation coefficient to obtain the structures of cognitive actions and then compare them between an expert and a novice designer. In many cases variance analysis (ANOVA) was used to carry out testing and comparisons among different sets of protocol data to confirm findings.

The analysis and the interpretation of design protocol are heavily reliant on statistical methods. Recently, information theory has been applied in statistical inference (MacKay 2003). Although it was proposed to model qualitative data in the 1980s (Krippendorff 1986), it has not been used in protocol analysis. Exploring the application of information theory may provide new insights into design protocol analysis.

Time Line of Design Activities

Many researchers have observed that the design activities change during a design session. Goldschmidt (1995) divided design sessions into episodes and reported the differences in the interconnectivity of ideas in those episodes. Gero and McNeill (1998) found that designers spent more time reasoning about the function and behaviour at the beginning of a session and spent more time reasoning about the structure towards the end of a session. Time series analysis in the protocol study of designers is rare. In other fields, such as the behavioural sciences, sequential analysis has been used to model interaction patterns (Gottman and Roy 1983; Bakeman and Gottman 1997).

Summary

In summary, the unit of analysis featured in protocol studies of designers varies because it is determined by the aim of the study (Hay et al. 2016). Without the same unit of analysis, there is no standardisation of the coding scheme or segmentation. This makes it difficult to compare the results of different studies, even with the same set of data. Most of the coding schemes are unique to the protocol data and cannot be reused in new circumstances, which impedes the accumulation of knowledge in this field. The interpretation of design protocol is heavily based on simple statistical measurements of the quantity of encoded data.

2.1.3 Linkography

Linkography takes a very different approach than other protocol analysis methods. Goldschmidt (1990) introduced linkography to protocol analysis. Briefly stated, instead of classifying the segments, it studies the interconnection among the segments. This approach considers both the content and the process. The segments are treated as "design moves." The definition of a design move together with an example of constructing a linkograph is provided in the next chapter, Sect. 3.1. Latter, Goldschmidt and Tatsa (2005, p. 595) stated:

> Linkography is based on the premise that effective reasoning in a creative endeavour must perforce aim at first mining and then relating to one another the many items of data that are relevant to the task.

A linkograph is constructed by discerning the relationships among the moves to form links. It can be seen as a graphical representation of a design session that traces the associations of every design move. The design process can then be examined in terms of the patterns in the linkograph that display the structure of design reasoning. Three distinct patterns were identified: chunk, a group of moves that are almost exclusively linked among themselves; web, a large number of links among a relatively small number of moves; and sawtooth track, a special sequence of linked moves. Goldschmidt also identified two types of links, namely, backlinks and forelinks. Backlinks are links of moves that connect to previous moves and forelinks are links of moves that connect to subsequent moves. The next chapter explains their respective differences.

Analysis of Linkographs

The progress of a design session is made observable through the analysis of linkographs. The analyses of chunk, web and sawtooth patterns is conducted qualitatively. Linkography has been used to investigate the structure of design idea generation processes and to compare design productivity (Goldschmidt 1990, 1992, 1995). The primary quantitative comparison in these studies was by link index and critical moves. Link index, also known as link density, refers to the number of links divided by the number of moves. Critical moves are design moves that are rich in links, usually more than five links. The combined critical moves of a sequence describe its critical path. Goldschmidt used these numbers and the critical path to benchmark the productivity of a design session.

Applications of Linkographs

Goldschmidt's linkography was used by van der Lugt (2003) to trace the design idea generation process and to correlate the creative qualities of ideas with the degree of their integration. He conducted four experiments, idea generation sessions under different conditions, and asked participants to appraise the ideas after the sessions. He extended linkography by identifying the link types in three categories: supplementary, modification or tangential links that correspond to small alterations of ideas, the same direction of ideas, or a different direction of ideas respectively. He found that a well-integrated creative process has a large network of links, a low level of self-links, and a balance of link types.

Dorst (2003) used linkographs to trace the linking behaviour of designers with regard to design problems and design solutions to reveal the reflective practice of designers.

Study of Design Teams by Linkography

Linkography was used to study design teams by Goldschmidt (1995) and van der Lugt (2003). Goldschmidt (1995) compared the productivity of the design processes of an individual and a team. Participants were asked to design a bicycle carrier for a backpack. The team consisted of three designers and conversational turn-taking was used to segment protocol; an utterance by one of the designers was defined as one move. Critical moves analysis and link index provided a quantitative means to compare the "solo design" and the "team effort." The experiments conducted by van der Lugt (2003) consisted of five advanced product design students. Linkographs were generated to study the effects of sketches in idea generation meetings. These studies indicate that this technique is not dependent on the number of participants.

2.2 Design: Process or Artefacts

This section presents a particular view of design research. A design (a noun) is usually described as a set of decisions that determines the relationships among geometries, materials and performance. Although there are different variables in different domains, the central activities of designing (a verb) remain very similar. The authors assume that these activities, notwithstanding some claims to the contrary, are scientifically observable. They include thinking and knowing (Cross 2007a), free-hand sketching and interactions (Lawson and Loke 1997; Schön and Wiggins 1992), the social construction of design solutions (Minneman 1991) and designing-by-making (Jones 1970). Some activities are harder to observe than others are.

This book is not about the exploration of design methods; rather, it explores the methods that can be used to study design activities. Certain views of designing will affect how studies are conducted. For example, within the information processing model of designing, some have placed more stress on the internal and external representation of information (Eastman 2001), whereas others emphasise the interaction of information (Gero and Kannengiesser 2000). These two different views have fostered their own distinct research streams. For example, Badke-Schaub et al. (2007) followed the former model to assess the development of shared representations in a design team, whereas Suwa et al. (2000) followed the latter model to investigate situated inventions and unexpected discoveries.

This book considers that designing involves some acts of manipulation of available material with knowledge to fulfil a set of requirements by imbuing them with appropriate qualities. The set of requirements may change during designing. These acts of manipulation are considered to be transformative processes. The available material can be viewed through the lenses of function, behaviour and structure. Chapter 3, Sect. 3.4, provides a detailed explanation. Appropriate qualities can be evaluated subjectively as well as by measurement.

2.2.1 Design Ideas

It is commonly held that good design ideas are essential for good design outcome. Linkography studies have suggested that good ideas have many interconnections with other ideas. Chapter 3, Sect. 3.3.2 challenges the supposition that more links are better by arguing that if there are too many links the ideas will be too similar, leading to less interesting designs. This might also indicate early fixation. Although Cross (2007a) suggested that fixation is not necessarily undesirable in the study of expert designers, the authors speculate that novel ideas are desirable in designing. It is conjectured that the study of a linkography may review the processes of a good design. Gero (2010) used linkography and some of the techniques described in this book to measure fixation and commitment while designing.

2.2.2 Design Process and Design Outcome Assessment

A design is generally assessed according to its outcome. The assessment criteria might vary in different domains. However, creativity remains one of the most important criteria. Other criteria include usability, aesthetic appeal, practicability, performance and functionality. However, what are the factors in the design process that constitute a good design?

Studies have shown that there are behavioural differences between design experts and novices during designing (Kavakli and Gero 2002, 2003; Cross 2004). Educators will benefit from knowing which design processes will yield desirable design outcomes. Yukhina (2007) examined the effects of cognitive abilities and learning styles on design students' academic performance. She found that visualisation abilities are the best predictors of academic performance. However, her study did not find any consistent correlation between the design processes (using protocol analysis with Gero and McNeill's (1998) coding scheme) and learning styles or cognitive abilities. As mentioned in Chap. 1, there have also been studies that have attempted to find the relationship between design performance and the integration of ideas and design strategies (Goldschmidt 1992; Kruger and Cross 2006). They too found no compelling evidence to directly associate the design process with the design outcome.

2.3 Conclusions

This brief review suggests that the quantitative methods for analysing design protocols are the primary tool to study the design process. Descriptive statistics and correlations are the dominant tools for analysis. We propose an exploration of other methods for analysis, such as information theory and Markov chains.

Many coding schemes have been developed for use with design protocols. They tend to be based on particular views of the activity of designing. Many of them are unique to the data to which they are applied. This limits the applicability of the results obtained. Where more general coding shemes have been attempted, they lacked sufficient generality that would make it feasible to reuse them in widely varying circumstances. Linkography, on the other hand, does not have any coding scheme and has been successfully used in studies of team designing activities. This study seeks to determine the potential for an extension of linkography as an a analysis tool.

Chapter 3
Theoretical Framework

This chapter introduces a generic approach to carry out protocol analyses of designers using the Function-Behavior-Structure (FBS) ontology. It suggests codifying design protocols into FBS design issues and deriving FBS design processes using two models, a syntactic model and a semantic model. The syntactic model assumes that any design issues is cognitively related to its immediately preceding issue and as a consequence there is a design process; the concept of using Markov analysis is also presented as a tool to examine the syntactic model. Semantic design processes are derived from ontologically coded linkographs. The construction of the linkograph is further examined in this chapter, as it is the foundation upon which further concepts are built. The information captured in the linkograph is studied using statistics and clustering. The rationale of using information theory, entropy, to measure the linkograph is presented, as well as a concise explanation of information theory.

3.1 Design Ontology and Ontological-Based Coding

This section explores the use of the FBS ontology (Gero 1990) to develop a general coding scheme. Its aim is to capture semantic information from design protocols. This semantic information can then be utilised: (1) to explore different aspects of designing according to the focus of interest; (2) to quantify the use cognitive resources; and (3) to locate different types of design transformation processes.

J.W.T. Kan and J.S. Gero, *Quantitative Methods for Studying Design Protocols*, DOI 10.1007/978-94-024-0984-0_3

3.1.1 FBS Ontology and Coding

The FBS framework (Gero 1990) models designing in terms of three classes of ontological variables: function, behaviour and structure. In this view the goal of designing is to transform a set of functions into a set of design descriptions (D). The function (F) of a designed object is defined as its teleology; the behaviour (B) of that object is either derived (Bs) or expected (Be) from the structure, where structure (S) represents the components of an object and their relationships. A design description is never transformed directly from the function but undergoes a series of processes among the FBS variables. These processes include: formulation which transform functions into a set of expected behaviours; synthesis, wherein a structure is proposed to fulfil the expected behaviours; an analysis of the structure produces derived behaviour; an evaluation process acts between the expected behaviour and the behaviour derived from structure; and documentation, which produces the design description. Based on the structure there are three types of reformulation: reformulation of structure, reformulation of expected behaviour and reformulation of function. Reformulation of function is relatively rare, as it changes or redefines the design problem. Figure 3.1 shows the relationships among the eight transformation processes and the three basic classes of variables. The problem space and solution space are expanded by the introduction of new

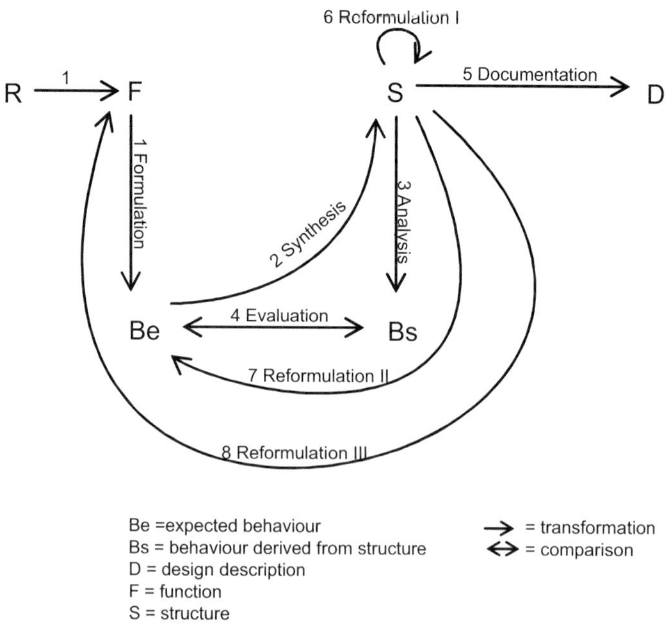

Fig. 3.1 The FBS ontology of designing

variables. These variables are introduced in the reformulation processes; structure, behaviour and function can all be part of reformulations.

The proposed generic coding scheme consists only of the function (F), expected behavior (Be), behavior derived from structure (Bs), structure (S), documentation (D) and requirement (R). Documentation and requirement are both describable in terms of function, behaviour or structure and do not require an extension of the FBS ontology. The protocols are segmented strictly according to these six categories. See Gero and Kannengiesser (2014) for a fuller explanation of the FBS ontology. Part of Gero and McNeill's (1998) coding scheme concerns the designer's reasoning about function, behaviour or structure in the problem domain. They do not separate the expected and derived behaviour.

3.1.2 *Situated FBS Ontology*

A number of new concepts constitute the situated FBS framework: the notion of situated cognition introduced by Clancey (1997); the idea of constructive memory based on the work of Dewey (1896) and Bartlett (1932); and the observation of designing as an "interaction of making and seeing" by Schön and Wiggins (1992). Gero and Kannengiesser (2002, 2004) developed these ideas further and integrated them into the FBS ontology to form the situated FBS framework by introducing interactions among three worlds—the external, interpreted and expected worlds. A brief description is provided here, however, for a complete exposition readers should consult the original papers (Gero and Kannengiesser 2002, 2004). A designer interacts and understands the external world through her/his interpretation of the external world to form memories of her/his interpreted world in terms of the FBS variables. In order to change the external world (the act of designing) s/he "focuses" to transform experiences to produce the expected world (also in terms of FBS) before taking action in the external world. In this framework the original eight processes are increased to twenty to allow for these additional activities.

Figure 3.2 presents the situated FBS ontology of designing. In the figure, R represents the requirement which is being interpreted in terms of function (F^i), behaviour (B^i), and structure (S^i). In the following text this interpretation process is represented by the symbol "↺". In the interpreted world there are four types of processes that the FBS variables can go through: transformation, represented by "→"; comparison, represented by "↔"; reflection or re-interpretation, represented by "↺"; and focusing, represented by "⇔". Focusing (⇔) refers to processes that produce an expected function (Fe^i) from an interpreted function (F^i), expected behaviour (Be^i) from interpreted behaviour (B^i), and expected structure (Se^i) from interpreted structure (S^i). Expected structure (Se^i) can also be transformed (→) from expected behaviour (Be^i), which in turn can be transformed from expected function (Fe^i), which represents the synthesis and formulation process in the original FBS

Fig. 3.2 The situated FBS ontology of designing

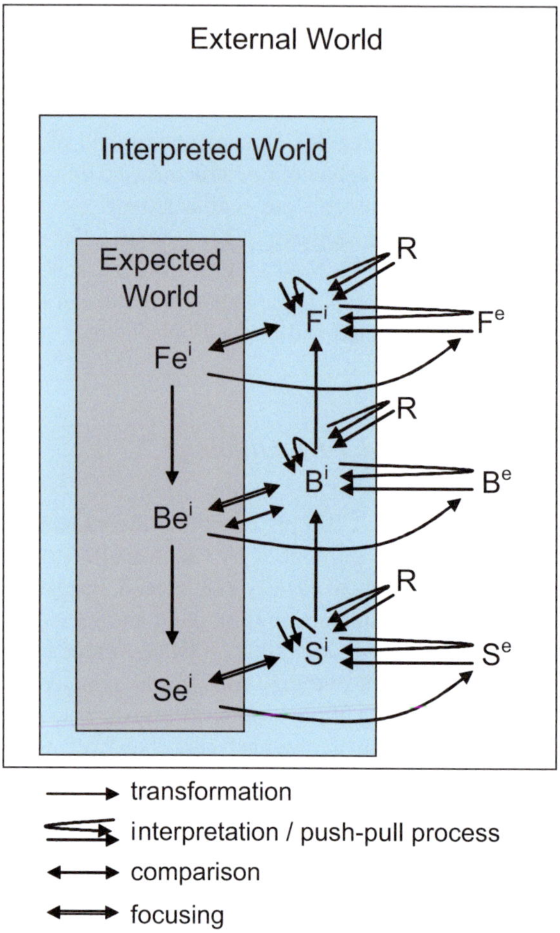

framework. The comparison (↔) is between expected behaviour (Be^i) and interpreted behaviour (B^i), which is similar to the evaluation in the original FBS framework.

Table 3.1 relates the twenty situated FBS processes to the original eight processes. Of particular interest are the formulation and reformulation processes in this framework. The formulation process involves: the interpretation of requirements (R) in terms of F^i, B^i, and S^i representations ($R \circlearrowleft F^i$, $R \circlearrowleft B^i$, $R \circlearrowleft S^i$); reflecting, based on experience, on those representations ($F^i \circlearrowleft F^i$, $B^i \circlearrowleft B^i$, $S^i \circlearrowleft S^i$, $F^e \circlearrowleft F^i$, $Be \circlearrowleft Bi$, $Se \circlearrowleft Si$); focusing on subsets of these internalised requirements ($F^i \Leftrightarrow Fe^i$, $B^i \Leftrightarrow Be^i$, $S^i \Leftrightarrow Se^i$); and the process $Fe^i \rightarrow Be^i$ that corresponds to the original formulation in the FBS framework. Focusing and reflecting ($F^i \Leftrightarrow Fe^i$, $B^i \Leftrightarrow Be^i$, $S^i \Leftrightarrow Se^i$) appear in all the three types of reformulations. The reformulations II and III are not limited to be driven by structure alone, but also by external representations of function ($F^e \circlearrowleft F^i$) and behaviour ($B^e \circlearrowleft B^i$).

Table 3.1 Situated FBS processes in relation to the eight FBS processes

1. Formulation	$R \circlearrowleft F^i$, $R \circlearrowleft B^i$, $R \circlearrowleft S^i$, $F^i \circlearrowleft F^i$, $B^i \circlearrowleft B^i$, $S^i \circlearrowleft S^i$, $F^i \Leftrightarrow Fe^i$, $B^i \Leftrightarrow Be^i$, $S^i \Leftrightarrow Se^i$, $Fe^i \rightarrow Be^i$
2. Synthesis	$Be^i \rightarrow Se^i$, $Se^i \rightarrow S^e$
3. Analysis	$S^i \rightarrow B^i$
4. Evaluation	$B^i \leftrightarrow Be^i$
5. Documentation	$Se^i \rightarrow S^e$, $Be^i \rightarrow B^e$ (optional), $Fe^i \rightarrow F^e$ (optional)
6. Reformulation I	$S^i \Leftrightarrow Se^i$, $S^i \circlearrowleft S^i$, $S^e \circlearrowleft S^i$
7. Reformulation II	$B^i \Leftrightarrow Be^i$, $S^i \rightarrow B^i$, $B^e \circlearrowleft B^i$, $B^i \circlearrowleft B^i$
8. Reformulation III	$F^i \Leftrightarrow Fe^i$, $B^i \rightarrow F^i$, $F^e \circlearrowleft F^i$, $F^i \circlearrowleft F^i$

$\rightarrow$ = transformation
$\leftrightarrow$ = comparsion
$\Leftrightarrow$ = focusing
$\circlearrowleft$ = interpretation, push-pull process

In protocol studies, designers can only be observed from the external world, the interpreted and expected world are all internal to the designers and can only be inferred from their protocols. Their actions are interpreted by the coder, so there is a degree of subjectivity in the analysis.

An example will be given later in this chapter to elaborate the situated FBS and FBS coding scheme. Before that, in the next section, an application of the notion of a general coding scheme for designing will be presented to show its potential contribution towards design research.

3.2 Meta-Analysis of Design Protocols Based on FBS Ontological Coding

As the FBS ontological coding scheme is a general coding scheme, it is possible to do meta-analysis of design protocols across domains. In order to investigate the commonalities across design domains, Gero et al. (2014a) proposed using the cumulative occurrence of design issues as a basis to examine the relative cognitive design effort across a design session. Cognitive design effort refers to the cognitive activities associated with designing. We will shorten the term to cognitive effort in this book. The cumulative occurrence of design issues models the cumulative cognitive effort across that design session. The cumulative occurrence of a design issue across all segments in a design protocol is calculated as follows: the cumulative occurrence (c) of design issue (x) at segment (n) is $c = \sum_{i=1}^{n} x_i$ where (x_i) equals 1 if segment (i) is coded as (x) and 0 if segment (i) is not coded as (x). Plotting the results of this equation on a graph with the segments (n) on the

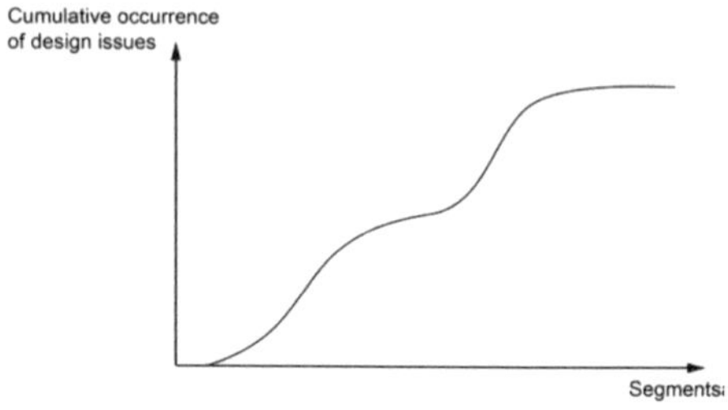

Fig. 3.3 Exemplary graph representing the cumulative occurrence of design issues (original caption of Gero et al. 2014b)

horizontal axis and the cumulative occurrence (c) on the vertical axis yields a visual representation of the cumulative cognitive effort represented by the occurrence of the design issues in a protocol, Fig. 3.3 (Gero et al. 2014a).

Based on the notion of cumulative occurrence of design issues, Gero et al. (2014a) utilised the following qualitative measures for each of the six classes of design issues:

- First occurrence at start: Which design issues first occur near the start of designing, and which first occur later?
- Continuity: Which design issues occur throughout designing, and which occur only up to a certain point?
- Shape of the graph: For which design issues is the cumulative occurrence graph linear, and for which is it non-linear?
- Slope: This is a measure for the speed at which design issues are generated.
- R^2 (coefficient of determination): This is a measure for the linearity of the graph. We will set a minimum value of 0.950 as a condition for linearity.

All of the above measures are independent of the length of the design session, which allow comparison of design protocols with different numbers of segments.

Gero et al. (2014a) used this model to examine 13 sets of design protocols drawn from a variety of studies carried out by different researchers in different countries involving design task and different levels of expertise. They found some commonalities, which is not surprising, given existing assumptions, observations and hypotheses about designing. For example, the design process commences with clarifying a set of requirements and functions that was shown by the discontinuous graphs of these two issues. However, some of the findings are unexpected and might bring new insight to the theory of designing. For example the structure issues occur continuously throughout design sessions and occur at a linear rate, Figs. 3.4 and 3.5 show the cumulative occurrence of structure issues of mechanical engineering students and a professional design team respectively. The results showed this linear cumulative occurrence of structure issues in all the 13 sets of design protocols. This is contrary to the notion that a design process start with requirements and functions and then, at a later stage, ends with structures and descriptions.

Gero et al. (2014b), using the same notion of cumulative occurrence of issues, examine three sets of data; the first set of data contains 18 design protocols of mechanical engineering students at different stages in design education; the second

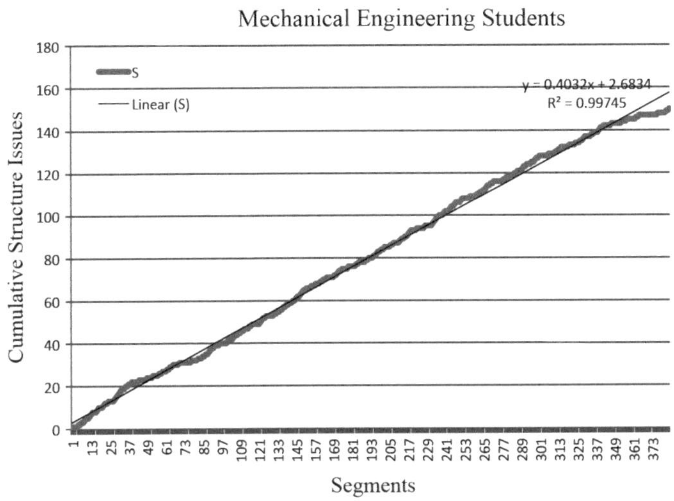

Fig. 3.4 Cumulative occurrence of structure issues of mechanical engineering students (after Gero et al. 2014a)

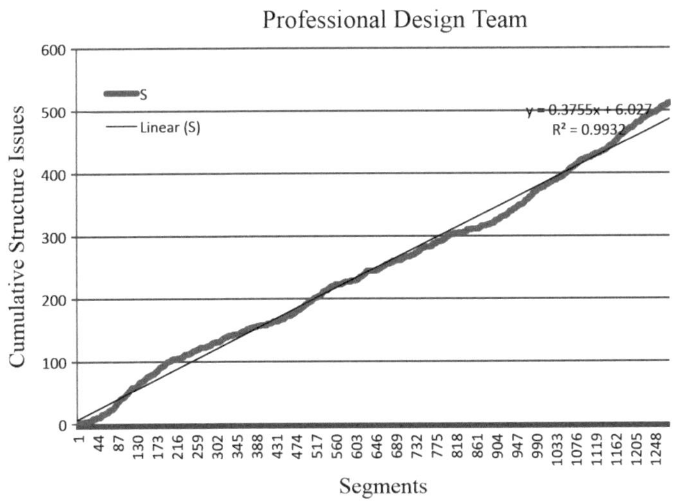

Fig. 3.5 Cumulative occurrence of structure issues of a professional design team (after Gero et al. 2014a)

set of data contains 31 design protocols of mechanical engineering students being taught different concept-generation methods; and the third set of data contains 42 sessions of software engineering students. Their results provide evidentiary support of the linearity of structure issues in all three datasets and in that there is no statistically significant differences between the mean slopes of the linear graphs. This implies that the cognitive effort expended on structure is expended uniformly across all the design sessions independent of the domain and task. Further, the rate of expenditure is independent of domain and task for these three studies of mechanical engineering students.

Kannengiesser et al. (2015), in a longitudinal study, using the same meta-analysis notion tested if design cognition of high school students who have taken pre-engineering courses (experiment group) would be different to those who have not (control group). Again, they found the same linearity of structure issues for the two groups.

The aim of this sub-section is to demonstrate the potential and significance of a general coding scheme so results of the studies described above are not fully described here.

3.3 Linkography

Linkography was first introduced to protocol analysis by Goldschmidt (1990) to assess the design productivity of designers. The design protocol is broken down into small units called "design moves". Goldschmidt defined a design move as "a step, an act, an operation, which transforms the design situation relative to the state in which it was prior to that move" (Goldschmidt 1995, p. 195) or "an act of reasoning that presents a coherent proposition pertaining to an entity that is being designed" (Goldschmidt 1992, p. 72). The following example is taken from Goldschmidt (1992). The four moves below are accompanied by a sketch, illustrated in Fig. 3.6.

1. If I look at the form again (it might also be the influence of having done entry 1, but) it seems that spatially, these are the larger directions (w, x).
2. I am getting one, two, three spaces here (p) and one, two (q) there.

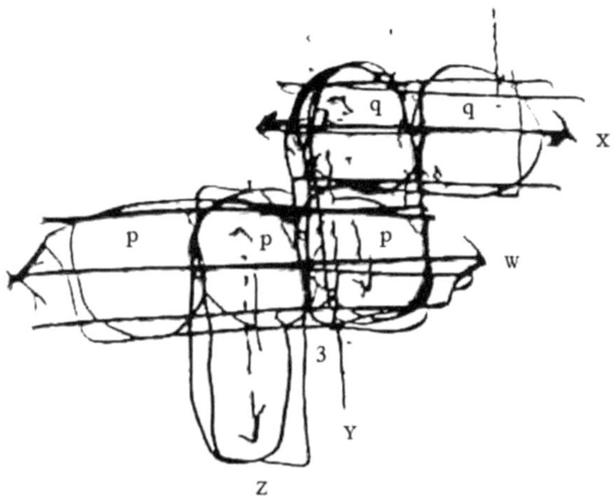

Fig. 3.6 Sketch from a protocol. Goldschmidt (1992), original caption

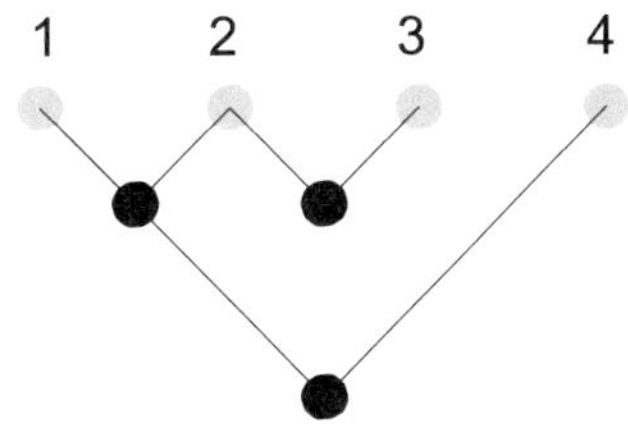

Fig. 3.7 Goldschmidt example of linkograph with four moves with the addition of nodes (shown in *black*) used in subsequent quantitative analyses

3. They're about square, so there is a tendency to try and see them as spaces.
4. These are secondary directions within the space (y, z), so the entry (3) is actually moving in along the secondary directions.

A linkograph is then constructed by linking related moves. The links are established by discerning, using domain knowledge and common sense, whether a move is connected to the previous moves. In her exposition, move 2, if judging from the verbalisation only, is not linked to move 1, but looking at the sketch, "the spaces it specifies are articulated through encircling the large directions of move 1" so move 2 is linked to move 1. Move 3 elaborates on move 2 but does not seem to link with move 1. Move 4 "discontinues the spatial diagnosis of moves 2 and 3" and returns to "the question of directions which was first brought up in move 1". Figure 3.7 is constructed by joining the linked moves. It can be seen as a graphical network of associated moves that represent the design session.

The design process can then be examined in terms of the patterns of move associations. Goldschmidt identified two types of links: backlinks and forelinks. Backlinks are links of moves that connect to previous moves. Forelinks are links of moves that connect to subsequent moves. In Fig. 3.2, moves 2 and 4 are backlinked to move 1 and move 3 is backlinked to move 2; move 1 is forelinked to moves 2 and 4, and move 2 is forelinked to move 3. Conceptually forelinks and backlinks are very different. Goldschmidt (1995, p. 196) stated, "backlinks record the path that led to a move's generation, while forelinks bear evidence to its contribution to the production of further moves."

The link index and critical moves were devised as indicators of design productivity. A link index is the ratio between the number of links and the number of moves; in this case it is 3/4. Critical moves are design moves that are rich in links. Figure 3.8 is another linkograph from Goldschmidt (1995), with six critical moves. They can be forelinks (moves 21, 26, and 28), backlinks (moves 15, 31, and 32), or both (moves 26 and 31). In her exposition, design productivity is positively related to the link index and critical moves; a higher value of link index and critical moves indicates a more productive design process. Later, Goldschmidt and Tatsa (2005) provided empirical evidence that quality outcome and creativity hinge on good ideas or what she called critical moves.

With an understanding of the construction of a linkograph, one is able to comment on the design behaviour without looking at the design protocol. Goldschmidt (1992) suggested that the linkograph pattern of productive designers will be different from that of less productive designers. Productive designers will

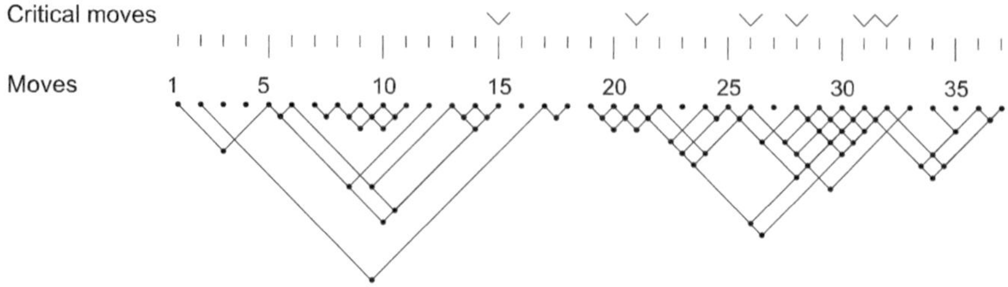

Fig. 3.8 Linkograph from Goldschmidt (1992); "v" indicates critical moves

elicit moves that have a high potential for connectivity to other moves, while less productive designers will exhibit random trails with moves that do not have high potential for contribution to the design concept. In addition, designers who start the design process by exploring different options and then select one to develop will produce a very different linkograph compared to designers who use a holistic approach without exploring different options.

3.4 Syntactic Design Process

We define syntactic design processes as the transformation of cognitively related design issues by assuming that any design issue is related to its immediately preceding issue. As we can see from the above subsections on linkography this assumption is not necessarily the case. In the Sect. 3.5, we define semantic design processes as design processes that are derived by considering the semantic linkage of design issues. We propose using the syntactic design processes as an efficient way to link design issues to produce design processes as it does not involve the labour intensive construction and arbitration of linkographs.

3.4.1 FBS-Based Design Issues of an Episode

If the FBS ontological coding scheme is used to study the issues of the four-move example depicted in Sect. 3.3, move 1 ("If I look at the form again it seems that spatially, these are the larger directions") will be coded as behaviour (Be), since it reasons the spatial behaviour of the form as directions. Move 2 ("I am getting one, two, three spaces here and one, two there") will be structure (S), because it describes spaces and their topology. Move 3 ("They're about square, so there is a tendency to try and see them as spaces") will be behaviour (Bs), as it rationalises and analyses the behaviour of the squares. Move 4 ("These are secondary directions within the space, so the entry is actually moving in along the secondary directions") will be behaviour (Be) again, as this move concerns the directional aspects of the spaces and the entrance. The sequence of the design issues is Be, S, Bs, and Be.

3.4.2 Syntactic Design Process of the Episode

In this book, processes are derived either syntactically or semantically. With the four design issues here we will have three syntactic processes. Be–S, which is a synthesis process, S–Bs, an analysis process and Bs–Be, which is considered an evaluation process.

3.4.3 Situated FBS-Based Design Issues of an Episode

Again, Goldschmidt's linkograph example in Fig. 3.6 is used to expound how the situated FBS framework can be used as a coding scheme. As each of the four moves contains more than one category of design issue, for example, accompanied with drawing actions, they have to be re-segmented. To avoid confusion, the new segments are called segments instead of moves. The 10 coding categories correspond to situated FBS framework are:

1. R: given requirements or derived from the brief
2. F^i: interpreted function either derived from requirements or ascribing meaning to the depicted structure
3. F^e: external representation of function, usually expressed in written words
4. Fe^i: expected function resulting from focusing on the interpreted function
5. B^i: interpreted behaviour from the depicted structure or requirements
6. B^e: external representation of behaviour, in terms of symbols or written words
7. Be^i: expected behaviour derived from the expected function or interpreted behaviour which result(s) from the requirements or interpreted structure
8. S^i: interpreted structure either from the external structure or from requirements
9. S^e: depiction that indicates the structure
10. Se^i: expected structure from expected behaviour or by focusing on the interpreted structure, sometimes without depiction

Assuming the interpretation of the sketches prior to the verbal protocol is segment 1 and is coded as S^i, the original move 1 is triggered by the designer looking at the sketch (form) again (re-interpreting the structure of segment 1) and then drawing the horizontal axis/direction *w*, *x*, which is a documentation of behaviour. This move is broken down into three segments:

2. "If I look at the form again it seems that spatially…"

This segment is a re-interpretation of behaviour of existing drawings, so it is coded as B^i and it is connected to segment 1.

3. "…these are the larger directions…"

This is coded as Be^i, because the designer is expecting the spatial behaviour of the structure in terms of direction. This segment is related to both segments 1 and 2.

4. *(Draw w x lines)*

This action is coded as S^e because it is a depiction of the spatial relationship, the topology of the structure. When considering the connections, this action is linked to the expected behaviour (previous segment 3) and the topology of the previous structure depiction (segment 1). This segment can be refined into two separated actions of depicting behaviour, but in this illustration there is no point in doing so, because *w* and *x* have not been referred to individually.

Move 2 is about reading the structure as five spaces, which can be further refined into five separate actions of depicting the structure (circling the 5 squares). Again, it is not separate and is truncated into two segments, one expectation of structure and one drawing action.

5. "I am getting 1, 2, 3 spaces here and 1, 2 there"

This segment is a re-interpretation of the depicted structure, which is related to the interpreted structure in segment 1 so it is coded as S^i, it is also related to segment 2 of "… look at form again… spatially" and the drawings in segment 1. This segment does not seem to relate to the directional aspect of segments 3 and 4.

6. *(Draw p and q squares)*

This segment is coded as S^e as it is about the form of the building, which is linked to the expected structure in segment 5 and the sketch in segment 1, because the sketch sets the boundary of the "spaces".

Move 3 is about justifying the spaces that the designer has just circled. This move is re-segmented into two segments, one interpretation and one expectation.

7. "They're about square…"

is coded as S^i, as it concerns the form. It is linked to segments 5 and 6, as the designer is focusing on *p* and *q* but not the overall form, so it is not linked to segments 1 and 2, nor does it link to segments 3 and 4 which are about axis and direction.

8. "so there is a tendency to try to see them as space"

is coded as Se^i, as the "… try to see them as space" is an expectation of the form. It is linked to segments 5, 6 and 7, as the state of affairs concerns the *p* and *q* spaces and not other things.

Move 4 returns to the directional aspect of the designed spaces and contains drawing actions. It also reflects on the influence of the axis towards the entrance. It has been divided into three segments.

9. "These are secondary directions within the space…"

This is a re-interpretation and expectation of the spatial behaviour of the design, so it is coded as Be^i. It is connected to segments 1, 2, 3 and 4 because they are all related to the orthogonal axis of direction.

10. *(Draw y z lines)*

As in segment 4, it is coded as S^e. It is a result of the above segment (9) and it hinges on previous depictions, so it is linked to segments 1 and 4.

Table 3.2 Segments coded with situated FBS issues

1	S^i	Structure before move 1
2	B^i	If I look at the form again it seems that spatially
3	Be^i	These are the larger directions
4	S^e	Draw w x
5	Se^i	I am getting 1, 2, 3 spaces here and 1, 2 there
6	S^e	Draw p and q
7	S^i	They're about square
8	Se^i	So there is a tendency to try to see them as space
9	Be^i	These are secondary directions within the space
10	S^e	Draw y z
11	B^i	So the entry (3) is actually moving along the secondary directions

11."…so the entry (3) is actually moving along the secondary directions."

This segment is coded as B^i, as the designer discovers the directional behaviour of the entrance. This seems to be only related to the idea of the secondary (y) axis the designer has just raised, so it is linked only to segments 9 and 10.

Table 3.2 shows the re-segmented protocol and their associated coded issues. This protocol does not contain examples of function or requirement coding categories.

3.4.4 Using Markov Chains to Describe the Design Process

This section uses the situated FBS-coded segments to illustrate some concepts of Markov analysis. In this analysis the links in linkographs are not considered; only the sequence of the design issues is used. Markov chains, also referred to as Markov analysis and Markov models, examine the sequence of events; they analyse or describe the probability of one event leading to another. In mathematics, a Markov chain is a discrete-time stochastic process with a number of states such that the next state solely depends on the present state. Markov chains have been used to analyse writers' manuscripts and to generate dummy text (Kenner and O'Rourke 1984); for the ranking of web pages by Google (Langville and Meyer 2006); and to capture music compositions and synthesise scores based on the analyses (Farbood and Schoner 2001).

In protocol analysis, McNeill et al. (1998), treating analysis, synthesis and evaluation as Markov states, found that the most likely event to follow analysis is a synthesis event. Also the most likely event after synthesis is an evaluation event and the most likely event after an evaluation event is a synthesis event.

To illustrate a Markov chain, each design issue of a segment is considered as an event. The purpose is to investigate the sequence of the events (coded segments) in relation to the probability of the previous events. The simplest Markov chain is the

Table 3.3 Percentage of one state to the next state

State	Next state				
	Be^i (%)	B^i (%)	Se^i (%)	S^e (%)	S^i (%)
Be^i				100	
B^i	100				
Se^i	50				50
S^e		33	33		33
S^i		50	50		

first-order chain which only examines the intermediate state after an event. In this example there are 11 segments, so there are 10 transitions from one state to another. Within the 11 segments there are only five different design issues being coded, B^i, Be^i, S^e, Se^i and S^i. Table 3.3 shows the percentages of each event in one state in relation to the next state. With this observation of events it can be turned into a transition matrix, Eq. 3.1, which represents a Markov process. The numbers in the matrix represent the probability of an event. Alternatively, this can be represented in graphic form, as illustrated in Fig. 3.9. This figure demonstrates that when the event B^i happens, the next event will be Be^i ($p(Be^i) = 1$). The event after Se^i will share a 50 % chance of being Be^i or S^i.

$$P = \begin{pmatrix} & Be^i & B^i & Se^i & S^e & S^i \\ Be^i & 0.00 & 0.00 & 0.00 & 1.00 & 0.00 \\ B^i & 1.00 & 0.00 & 0.00 & 0.00 & 0.00 \\ Se^i & 0.50 & 0.00 & 0.00 & 0.00 & 0.50 \\ S^e & 0.00 & 0.33 & 0.33 & 0.00 & 0.33 \\ S^i & 0.00 & 0.50 & 0.50 & 0.00 & 0.00 \end{pmatrix} \tag{3.1}$$

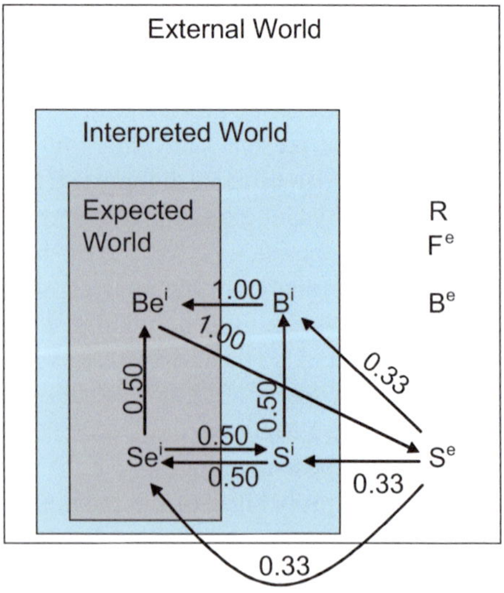

Fig. 3.9 The probability from one FBS state to another FBS state

3.4.5 Some Properties of Markov Chains

Kemeny and Snell (1960) classified three types of Markov chains based on their behaviours: absorbing, regular and ergodic. An absorbing chain is one that consists of states that, once entered, will never be left. This is not likely to happen in design protocols. For example if S is the absorbing state, once the sequence has reached this state, all the following states will be S and nothing else.

A chain is regular if and only if it is possible to be in any state after a number of steps no matter what the starting state is. A chain is ergodic only if it is possible to transit directly from a state to any other state. Once a system enters an ergodic set it will never leave it. Here, only the properties of regular chains will be considered.

Probability Matrix/Vector of Regular Chains

Some behaviours of a regular Markov chain are:

the powers of P^n approach a probability matrix A;
each row of A is the same probability vector $\alpha = a_1, a_2, \ldots, a_n$; and
for any probability vector π, $\pi \cdot P^n$ approaches the vector α as n approaches infinity.

$$\text{i.e. } \alpha P = \alpha \tag{3.2}$$

Essentially these mean that there is a limiting probability a_j of being in the state s_j independent of the starting state. Using the above example and the equation $\alpha P = \alpha$, P can be substituted by Eq. 3.6 to obtain the following five equations.

$$\begin{aligned}
&a_2 + \frac{1}{3}a_3 = a_1\\
&\frac{1}{3}a_4 + \frac{1}{2}a_5 = a_3\\
&a_1 = a_4\\
&\frac{1}{2}a_3 + \frac{1}{3}a_4 = a_5
\end{aligned}$$

Since α is a probability vector the sum of the elements equals one.

$$a_1 + a_2 + a_3 + a_4 + a_5 = 1$$

The unique solution to these equations is:

$$a = \left(\frac{1}{4}\frac{1}{6}\frac{1}{6}\frac{1}{4}\frac{1}{6}\right) \tag{3.3}$$

This means that for a large number of coded segments one can expect $\frac{1}{4}$ (a_1) of the segments will be Be^i, $\frac{1}{6}$ (a_2) of the segments will be B^i, $\frac{1}{6}$ (a_3) of the segments will be Se_i, $\frac{1}{4}$ (a_4) of the segments will be S^e, and $\frac{1}{6}$ (a_5) of the segments will be S^i. This distribution of design issues is a little different from the original distribution

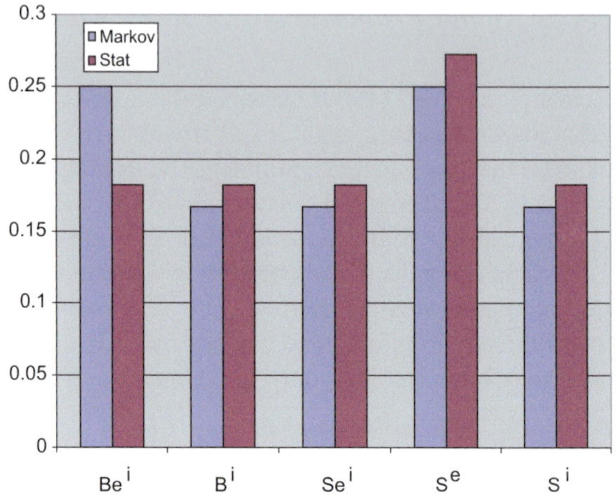

Fig. 3.10 Charts showing the code distribution probabilities based on Markov analysis and statistics

Fig. 3.10 shows the differences. The chart suggests that, when there is a large number of coded segments, the Markov analysis predicts more Be^i design issues than the statistical prediction, whereas the Markov prediction of the occurrence of other segments is lower than the prediction by statistical analysis.

Traditional protocol analysis is based heavily on statistical analysis: this contains the assumption that each segment is an independent event. Markov analysis, based on the probability of relationship with the last event, provides another venue for insight into the design activities.

First Passage Times

The mean first passage time is the average number of steps traversed before reaching a state from other states. The mean passage time can be obtained from the transition matrix and the probability matrix. Kemeny and Snell (1960) proved that the mean first passage matrix M is given by:

$$M = \left(I - Z + EZ_{dg}\right)D \tag{3.4}$$

where I is an identity matrix, E is a matrix with all entries of 1, D is the diagonal matrix with diagonal elements $d_{ii} = 1/a_i$, and Z is the fundamental matrix such that

$$Z = (I - (P - A))^{-1} \tag{3.5}$$

From Eq. 3.8

$$A = \begin{pmatrix} \frac{1}{4} & \frac{1}{6} & \frac{1}{6} & \frac{1}{4} & \frac{1}{6} \\ \frac{1}{4} & \frac{1}{6} & \frac{1}{6} & \frac{1}{4} & \frac{1}{6} \\ \frac{1}{4} & \frac{1}{6} & \frac{1}{6} & \frac{1}{4} & \frac{1}{6} \\ \frac{1}{4} & \frac{1}{6} & \frac{1}{6} & \frac{1}{4} & \frac{1}{6} \\ \frac{1}{4} & \frac{1}{6} & \frac{1}{6} & \frac{1}{4} & \frac{1}{6} \end{pmatrix} \tag{3.6}$$

Put Eqs. 3.6 and 3.1 into Eq. 3.5

$$Z = \begin{pmatrix} 0.69 & -0.04 & -0.04 & 0.44 & -0.04 \\ 0.44 & 0.79 & -0.21 & 0.19 & -0.21 \\ 0.10 & -0.10 & 0.90 & -0.15 & 0.24 \\ -0.06 & 0.13 & 0.13 & 0.69 & 0.13 \\ 0.02 & 0.18 & 0.18 & -0.23 & 0.85 \end{pmatrix} \quad (3.7)$$

Solving Eq. 3.4 the following matrix is obtained.

$$M = \begin{pmatrix} & Be^i & B^i & Se^i & S^e & S^i \\ Be^i & 4.00 & 5.00 & 5.67 & 1.00 & 5.33 \\ B^i & 1.00 & 6.00 & 6.67 & 2.00 & 6.33 \\ Se^i & 2.33 & 5.33 & 6.00 & 3.33 & 3.67 \\ S^e & 3.000 & 4.00 & 4.67 & 4.00 & 4.33 \\ S^i & 2.67 & 3.67 & 4.33 & 3.67 & 6.00 \end{pmatrix} \quad (3.8)$$

Thus, for example, if the designer is at the $\mathrm{Be^i}$ state, the mean number of steps before another $\mathrm{Be^i}$ state is 4, the mean number of steps before a $\mathrm{B^i}$ state is 5; before an $\mathrm{Se^i}$ state is 5.67; before an $\mathrm{S^e}$ state is 1; and before an $\mathrm{S^i}$ state is 5.33.

In this example, the shortest paths are $\mathrm{Be^i}$ to $\mathrm{S^e}$ and $\mathrm{B^i}$ to $\mathrm{Be^i}$ (1 mean step) and the longest one is $\mathrm{B^i}$ to $\mathrm{Se^i}$ (6.67 mean steps). Since this example has limited observations, only 11 states and transitions, the results. The authors do not attempt to interpret these numbers; they only serve as a demonstration of how Markov analysis can be used to study design protocols. One would expect it will take more steps to move from Function (F) to Structure (S) and fewer steps to move from expected Behaviour (Be) to Structure (S).

Statistical descriptions, cumulative occurrence, Markov chains and mean passage times provide quantitative models of design cognition—the cognitive behaviour of designers. They are used to gain insight into designing as a process. We now move on to how we can produce richer representations from protocol source data and how we can generate further quantitative models of design cognition that enhance our understanding of design. We can then use these models to examine similarities and differences in a large variety of design conditions.

3.5 Semantic Design Process

Semantic design processes are the design processes that are derived by considering the semantic linkage of design issues. After constructing the linkograph, if there are n links there will be n processes.

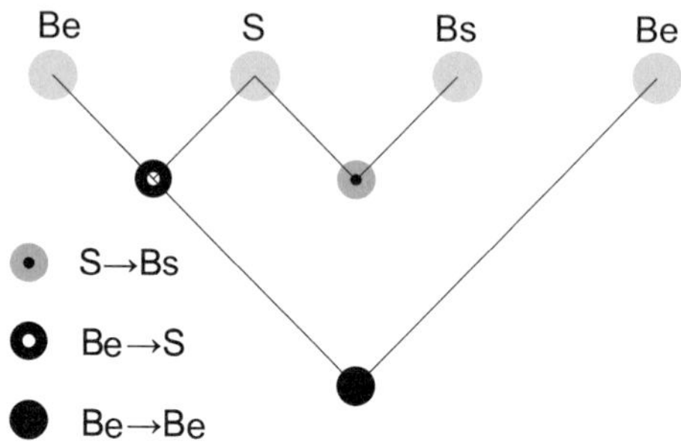

Fig. 3.11 Goldschmidt examples with coded FBS issues

3.5.1 *Deriving FBS Design Processes*

Using Goldschmidt's example of four move in Sect. 3.4 by combining Fig. 3.6 and the FBS ontology codes in Sect. 3.4.1 we get Fig. 3.11, which shows the linkograph together with the FBS issues. In this example, three processes are derived. The link from move 1 to move 2 (Be → S) meets the definition of synthesis. The link from move 2 to move 3 (S → Bs) meets the definition of analysis. These agree with the understanding of the protocol. The last process, from move 1 to move 4 (Be → Bs), should not be classified as evaluation if we examine the design protocol. Move 4 is triggered by the "direction" aspect the designer discovers in move 1. If the behaviour code (B) is to be separated into expected behaviour (Be) and behaviour is derived from structure (Bs), the first move and the fourth move should be coded as Be, as both moves were anticipating the directional behaviour of the design. However, there is no process within the FBS framework to describe the (Be → Be) process, which can be viewed as a reflection process. In order to better capture design information from the design protocol, the situated FBS ontology will be used.

3.5.2 *Deriving Situated FBS Design Processes*

Since we have already subdivided the protocols using the situated FBS coding scheme we can construct a new linkograph by discerning the connections among the segments in Table 3.2. Table 3.4 shows the rotated linkograph in relation to the segments and situated FBS issues. The reasoning of the connections among segments is too lengthy to depict here. Interested readers can examine Table 3.4 to see if they agree with the authors' discernment. There are 23 links in the linkograph. If each link is considered as a transformation process, there will be 23 processes. If segment n is connected to segment (n + i), it is represented by n ⇝ (n + i) without specifying the type of processes as in Table 3.5. It also shows the frequencies and links of the derived processes. Observing the table, there are more structure-initiated processes than behaviour-initiated processes, 16 against 7.

Table 3.4 Example of coding with situated FBS

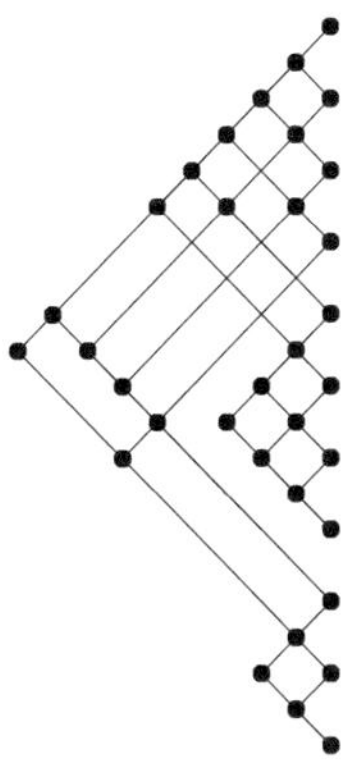

1	S^i	Structure before move 1
2	B^i	If I look at the form again it seems that spatially,
3	Be^i	These are the larger directions
4	S^e	Draw w x
5	Se^i	I am getting 1, 2, 3 spaces here and 1, 2 there
6	S^e	Draw p and q
7	S^i	They're about square
8	Se^i	So there is a tendency to try to see them as space
9	Be^i	These are secondary directions within the space
10	S^e	Draw y z
11	B^i	So the entry (3) is actually moving along the secondary directions

Table 3.5 Derived processes

Derived process	Frequency	Links between segments (segments inside brackets)
$B^i \rightsquigarrow Be^i$	2	(2 3) and (2 9)
$B^i \rightsquigarrow Se^i$	1	(2 5)
$Be^i \rightsquigarrow B^i$	1	(9 11)
$Be^i \rightsquigarrow Be^i$	1	(3 9)
$Be^i \rightsquigarrow S^e$	2	(3 4) and (9 10)
$S^i \rightsquigarrow B^i$	1	(1 2)
$S^i \rightsquigarrow Be^i$	2	(1 3) and (1 9)
$S^i \rightsquigarrow Se^i$	2	(1 5) and (7 8)
$S^i \rightsquigarrow S^e$	3	(1 4), (1 6), and (1 10)
$Se^i \rightsquigarrow S^i$	1	(5 7)
$Se^i \rightsquigarrow Se^i$	1	(5 8)
$Se^i \rightsquigarrow S^e$	1	(5 6)
$S^e \rightsquigarrow B^i$	1	(10 11)
$S^e \rightsquigarrow Be^i$	1	(4 9)
$S^e \rightsquigarrow S^i$	1	(6 7)
$S^e \rightsquigarrow Se^i$	1	(6 8)
$S^e \rightsquigarrow S^e$	1	(4 10)

It can be observed that there are 17 types of processes in this design session. Some of these derived processes map directly to the situated FBS processes. For example, in the first row of Table 3.10, the two $B^i \rightsquigarrow Be^i$ processes correspond to the focusing process ($B^i \Leftrightarrow Be^i$). In these two instances the designer focuses on the directional aspect (segment 2 to segments 3 and 9) of the spatial form.

However, some derived processes do not match the situated FBS processes. For example, in the fifth row of Table 3.5, the two $Be^i \rightsquigarrow Se$ processes cannot be found in Fig. 3.2. Revisiting the design protocol suggests these two occurrences of structure depiction (S^e) are depicting topological directions (*w, x, y* and *z*) and there should be a topological expectation of structure (Se^i) after the expected directional behaviour (Be^i). In this case, the $Be^i \rightsquigarrow S^e$ processes map to two situated FBS processes represented by $Be^i \rightarrow Se^i \rightarrow S^e$.

By re-visiting the protocol, codes are added in between the "mismatched" processes. For example, in a large number of cases the external structure code (S^e) requires a structure interpretation code (S^i). For examples in segment 11 the designer discovers that the entrance 3 is "moving along the secondary direction"; this is an analysis of the depicted structure of *y* and *z* (segment 10) and the entrance. Before s/he analyses the structure, an interpretation of the structure is required. In some cases two additional codes are required, for example, the link between the depicted structures ($S^e \rightsquigarrow S^e$) segment 4 and segment 10. In Segment 4 the designer draws the *w* and *x* axes and in segment 10 the designer draws the *y* and *z* axes. These two segments are linked because without drawing the major *x* and *w* directions, s/he might not discover the "secondary" directions *y* and *z*. In the situated FBS framework, this drawing action involves interpreting existing drawings, focusing on the directions and then expecting the "secondary" directions before drawing *y* and *z*. These three processes are represented by: $S^e \circlearrowleft S^i \Leftrightarrow Se^i \rightarrow S^e$.

Table 3.6 summarises the mapping of the derived processes to the situated FBS processes. Seven out of the 17 processes can be directly mapped to the situated FBS framework. Out of the 10 types of processes that cannot be directly mapped to the ontology, eight require one additional situated FBS state between the codes assigned to segments, two require two additional situated FBS states between the codes assigned to segments. These added codes can be seen as states (or cognitive activities) that are not directly observable.

3.6 Statistical Analysis

As seen in Fig. 3.7 and Table 3.5, using simple counting can help to compare and understand an episode. In this section we explore different statistical analysis methods to examine design protocols coded with the FBS ontology and linkography. Our goal is to provide a platform for quantitative descriptions of the cognitive activities related to designing.

3.6.1 Descriptive Statistics of Design Issues and Processes

Descriptive statistics of design issues can quantify the types of cognitive activities used during the episode. For example counting the different issues of Fig. 3.11 we

Table 3.6 Derived processes mapped to situated FBS processes

Derived process	Situated FBS process	Comments
B^iBe^i	$B^i \Leftrightarrow Be^i$	Focusing on the expected behaviour of "larger" and "secondary" axis, a kind of Reformulation II
$B^i\ Se^i$	$B^i \Leftrightarrow Be^i \rightarrow Se^i$	Expected behaviour added between behaviour "… it seems spatially" and expected structure "I am getting 1, 2, 3 spaces" to complete the synthesis process
$Be^i\ B^i$	$Be^i \leftrightarrow B^i$	Evaluation of axial behaviour of entrance 3
Be^i Bei	$Be^i \Leftrightarrow B^i \Leftrightarrow Be^i$	Interpreted behaviour of "direction" added between the expected behaviour of "larger" and "secondary" directions, a kind of Reformulation II
Be^i Se	$Be^i \rightarrow$ Sei $\rightarrow S^e$	Expected structure added to complete the documentation process
$S^i\ B^i$	Si $\rightarrow B^i$	Analysis of structure
$S^i\ Be^i$	$S^i \rightarrow$ Bi $\Leftrightarrow Be^i$	Interpreted behaviour of "direction" added before the expected behaviour of "larger" and "secondary" directions
S^i Sei	$S^i \Leftrightarrow Se^i$	Focusing on the shape and "square"
$S^i\ S^e$	$S^i \Leftrightarrow Se^i \rightarrow S^e$	Focusing on the directions and square space before depicting them
Sei Si	$Se^i \Leftrightarrow S^i$	Focusing on the interpretation of squares
Sei Se^i	$Se^i \Leftrightarrow S^i \Leftrightarrow Se^i$	Justifying interpreting the expected structure as expected spaces
Sei S^e	$Se^i \rightarrow S^e$	Documentation of the p and q spaces
$S^e\ B^i$	$S^e \circlearrowleft$ Si $\rightarrow B^i$	Analysis of axis entrance 3 and y, interpreted structure added
$S^e\ Be^i$	$S^e \circlearrowleft$ Si $\rightarrow B^i \Leftrightarrow Be^i$	Interpreted structure and interpreted behaviour of "direction" added before the expected behaviour of "larger" and "secondary" directions
$S^e\ S^i$	$S^e \circlearrowleft S^i$	Interpreting the p and q spaces as squares
$S^e\ Se^i$	$S^e \circlearrowleft S^i \Leftrightarrow Se^i$	Interpreted structures of space before expecting the structure as spaces, p and q
$S^e\ S^e$	$S^e \circlearrowleft S^i \Leftrightarrow Se^i \rightarrow S^e$	Interpreted x w direction and expected secondary direction before depicting y z

get Fig. 3.12. This can be understood as during this episode, 75 % of the cognitive effort was spent on behavioural issues and 25 % of the cognitive effort was spent on structure issues.

However, if using the finer grained situated FBS model to investigate the episode and counting the occurrences of the situated FBS issues, Fig. 3.13a can be obtained. 36 % of the cognitive effort were put into behaviour issues. This cognitive effort was divided equally between the interpreted world and the expected world. The remaining 64 % were of structure issues; three out of seven of those were external depiction. The remaining structure issues were again equally distributed between the interpreted world and the expected world. Figure 3.13b shows the

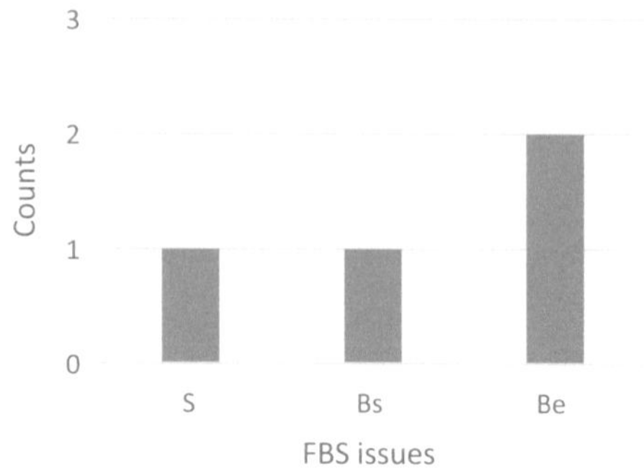

Fig. 3.12 Counting the FBS issues

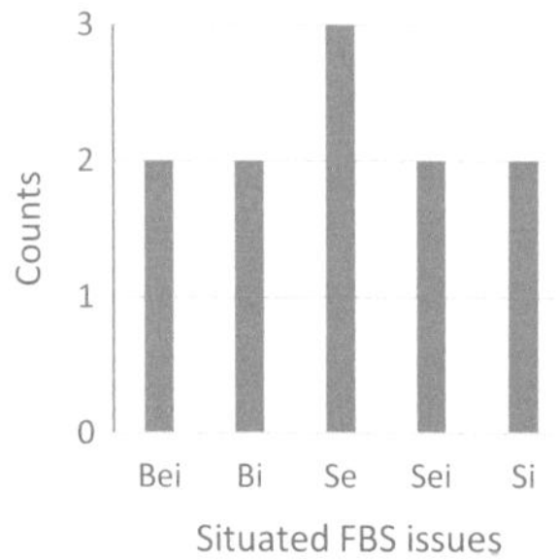

(a) Counting the situated FBS issues

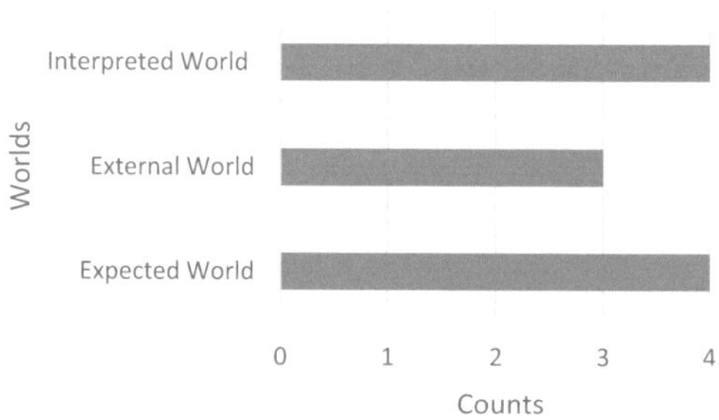

(b) Counting issues in different worlds

Fig. 3.13 Counting the situated FBS issues

count of design issues in relation to the three worlds. Descriptive statistical examination of a design episode with the situated FBS ontology will give addition insights into the cognitive effort expended in different areas.

Similarly, syntactic and semantic FBS design processes of a design session or a design episode can be counted. The application of descriptive statistics to protocols from experiments will be given in Chap. 4.

3.6.2 *Statistical Inference of Design Protocol: p Value*

Descriptive statistics summarize or quantitatively describe designing in terms of the FBS ontology. Inferential statistics can be used to test hypotheses about designing. To illustrate this we use a common statistical test called paired Student's *t* test. To put it into context, we construct an example from Bilda's (2006) mental imagery design experiment.

Student's t test is one of the most commonly used techniques for testing a hypothesis on the basis of a difference between sample means. A paired t test looks at the difference between paired values in two samples, takes into account the variation of values within each sample, and produces a single number. Statistical software reports results as a probability. This probability is called the p value. The p value is not produced directly by the t test, it is calculated in one further step, using the outcome of the t test. In another words, it determines a probability of the chance of the two populations are the same with respect to the variable tested. The p value gives a predictive answer to the question of how certain it is that the null hypothesis is true. The lower this value is, the less likely the difference is by chance. The p value helps one to decide whether or not to accept the null hypothesis. Typically in protocol analysis significance level of 0.05 is used as a cut-off point to reject the null hypothesis.

Back to our example, the idea behind Bilda's experiment is that when a designer does not have access to sketching in the early conceptual stage, it will affect both the design process and the design outcome. Design literature shows a common agreement that sketching is essential for conceptual designing. However, Toker (2003) documented that Frank Lloyd Wright could conceive and develop the entire design using imagery alone and produce an external representation at the end of the process. Blida's study aimed to investigate the effects of not having access to sketching, in the early conceptual design phase, on the cognitive behaviour of a designer. Data collection involved six expert architects working on two different design problems on the same site under two different conditions, one in which they were blindfolded and hence could not sketch (called the experiment condition) and one in which they were sketching (called the control condition). Further details of the experiment details will be given in Chap. 6.

Here, to illustrate the use of paired t test, we simplify the study by testing the hypothesis that the blindfolded sessions will have less cognitive activities after 20 min of the design session. The reasons behind this hypothesis are first of all, on average, our attention span is about 20 min and second the cognitive activities will slow down because of memory load and the unavailability of sketches to off load cognition. Table 3.7 shows the number of segments in the first 20 min and the rest of the session with respect to these two conditions. The assumption is the number of segment represents number of cognitive activities. The null hypothesis is there will

Table 3.7 Number of segments in the first 20 min and the rest of the session

	Blindfolded no. of segments			Sketch no. of segments		
	20 min	Rest	Total (45 min)	20 min	Rest	Total (45 min)
Architect 1	89	78	167	68	77	145
Architect 2	63	91	154	77	107	184
Architect 3	87	82	169	65	77	142
Architect 4	92	75	167	74	95	169
Architect 5	73	72	145	91	62	153
Architect 6	69	53	122	71	101	172

Table 3.8 Number of segments in the first 20 min and the rest of the session

	% segments BF last 25 min	% segments SK last 25 min
Architect 1	46.7	53.1
Architect 2	59.1	58.2
Architect 3	48.5	54.2
Architect 4	44.9	56.2
Architect 5	49.7	40.5
Architect 6	43.4	58.7
Average	48.7	53.5

be no difference in the percentage of cognitive activities of the two condition. Table 3.8 shows the percentages of segments in the last 25 min. Table 3.9 shows a typical summary of results when doing a t test on Table 3.8. The normal convention to report this results is: $t(5) = -1.34$, $p < 0.12$. Taking $p < 0.05$ to be the significance level, we cannot reject the null hypothesis, which is there is no difference in the percentage of cognitive activities of the two condition. Note that in this example we are using a number of participants that may be too small to produce statistical robustness, i.e., more reliable results will be produced with a larger sample size.

Does the result conclude there is no difference in the cognitive activities when designing blindfolded? No, the t test result suggest there is a difference of t value of −1.34, and at the probability of the difference being due to chance is less than 0.12, so we cannot be confident in rejecting the null hypothesis with this set of data. Any standard statistics textbook provides a detailed exposition of testing.

Table 3.9 Typical t test results

	% segments BF last 25 min	% segments SK last 25 min
Mean	48.72	53.48
Variance	30.97	45.06
Observations	6	6
Pearson correlation	−0.006013977	
Hypothesized mean difference	0	
df	5	
t Stat	−1.335499967	
$P(T \leq t)$ one-tail	0.119639738	
t Critical one-tail	2.015048373	
$P(T \leq t)$ two-tail	0.239279477	
t Critical two-tail	2.570581836	

3.6.3 Statistical Description of Linkographs

Classical protocol analysis uses statistics to measure segment categories. In linkography there is no categorical data. However, it can be observed in the graphs that some parts have a higher density of links than others. This section uses standard statistics and methods of clustering to describe a linkograph in such a way as provide information on which to base further insights into designing.

The linkograph in Fig. 3.8 can be re-represented by taking out all the linking lines, as in Fig. 3.14. Here the first move is assigned as the origin and there is a one-unit separation between each move. The position of each node (link) will have a coordinate in the X-Y plane. The linkograph can then be statistically described in terms of the total number of nodes and the statistical position of links, which are the mean values of $(\bar{x}, \bar{y})$ and their standard deviations (σx, σy). The total number of nodes indicates the level of saturation of a linkograph. Normalising this number against the number of moves will be the link index.

Table 3.10 and Fig. 3.15 show the statistics and scatter plot of the linkograph. A higher mean value of x, $\bar{x}$, implies that more links appear at the end of a session and a lower value suggests that more linked nodes are present at the beginning of the session. A higher mean value of y, $\bar{y}$, indicates longer linking lengths. However, the mean values do not include the dispersion of the distribution, therefore, standard deviations are measured to indicate how concentrated the nodes are clustered around the means. Tables 3.11 and 3.12 relate the appearance of linkographs, with the same number of links, to the statistical values of x and y respectively. The figures in the two tables are exaggerated for illustration. For this example, Fig. 3.12 and Table 3.10, there are more links towards the end of the session since $\bar{x}$ is greater than the median or middle point.

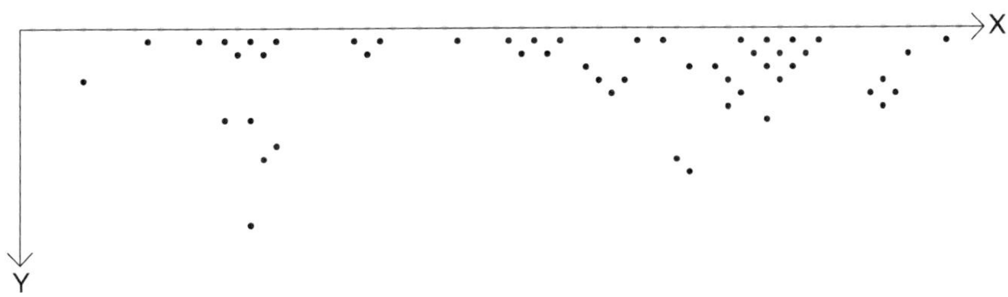

Fig. 3.14 Re-representation of the linkograph in Fig. 3.8 with nodes only in a 2-D space, the *lines* connecting the nodes to the moves have been removed, a *black dot* denotes a link between two moves

Table 3.10 Descriptive statistics of the example linkograph

	N	Minimum	Maximum	Mean $(\bar{x}, \bar{y})$	Std. deviation
X	52	3.00	36.50	22.01	9.41
Y	52	7.50	0.50	1.76	1.56

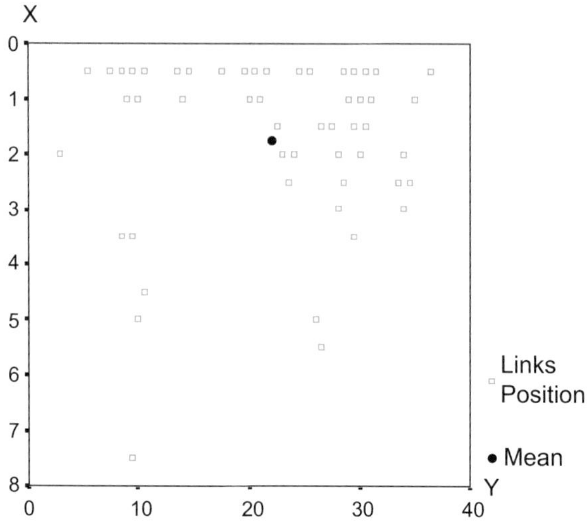

Fig. 3.15 Scatter plot of the linkograph with mean value

Table 3.11 Shape of a linkograph in relation to values of $\bar{x}$

X Axis	Small standard deviation	Large standard deviation
Small mean, $\bar{x}$	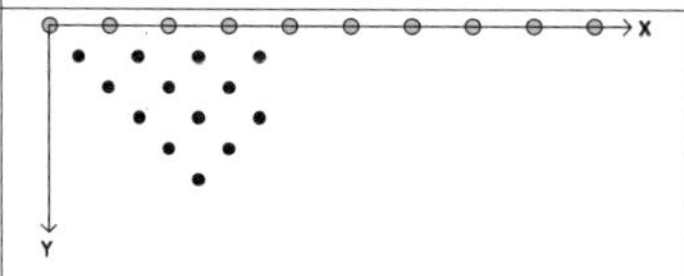	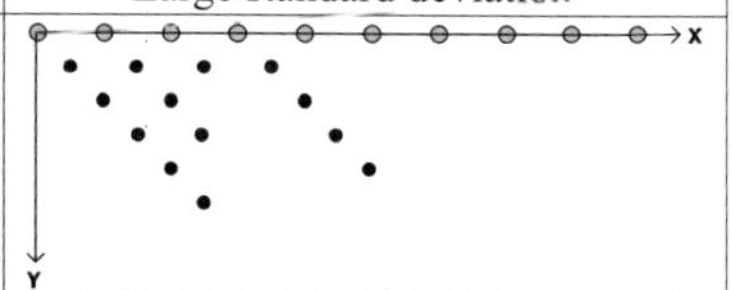
Large mean, $\bar{x}$	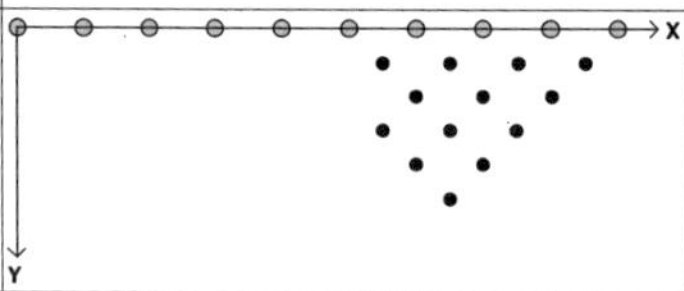	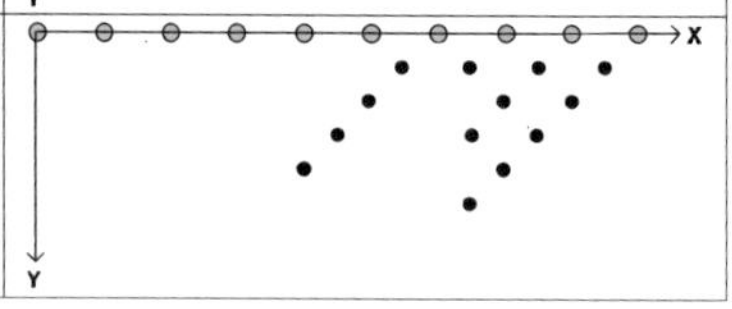

Table 3.12 Shape of a linkograph in relation to values of $\bar{y}$

Y Axis	Small standard deviation	Large standard deviation
Small mean, $\bar{y}$	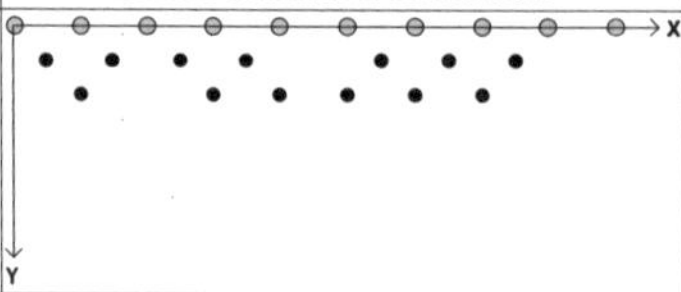	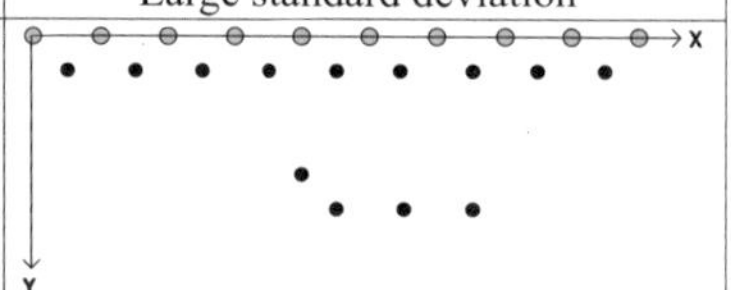
Large mean, $\bar{y}$	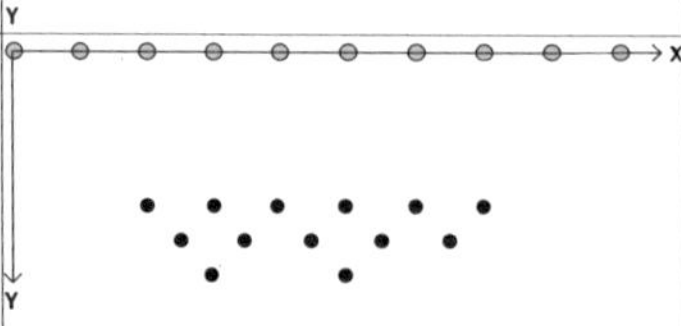	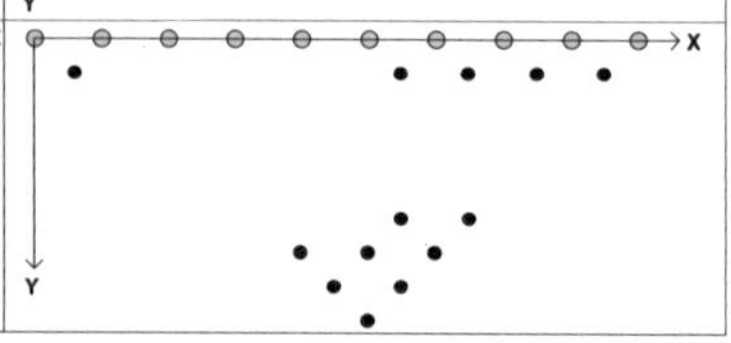

Cluster Analysis of Linkographs

Examining Fig. 3.8, there seem to be two chunks in this linkograph. The first chunk is from move 1 to move 18 and the second chunk from move 19 to 37. Comparing these two chunks in Fig. 3.8 to the scatter plot of Fig. 3.12, the links in a linkograph can be considered as data points that may form clusters in the x-y plane. These clusters resemble the chunks of ideas that are interlinked. Any clustering algorithm can be used to explore whether it is possible to cluster these two chunks automatically. This will complement the visual inspection to find the number of chunks and eliminate subjectivity. Most clustering algorithms can handle both continuous and categorical variables. The positions of links, those two-dimensional points (x, y) (nodes), are the data for clustering. In the first step of this procedure, the data are pre-clustered into many small sub-clusters, according to the selected criteria. Then, the algorithm clusters the sub-clusters that were created in the pre-cluster step into the desired number of clusters. If the desired number of clusters is unknown, the algorithm automatically finds the appropriate number of clusters according to the criteria. In this study the x and y variables were treated as continuous and Euclidean distance (the two-dimensional distance between links (x_i, y_i); and (x_j, y_j) computed by: $\sqrt{(x_i, y_i)^2 + (x_j, y_j)^2}$) was used to compute the distance among clusters. Akaike's information criterion (Akaike 1973), based on the maximum likelihood principle, was used for determining the number of clusters. Figure 3.16 shows the two groups of clusters found by the algorithm (here we used SPSS) which resemble the chunks. Table 3.13 shows the cluster distribution and Table 3.14 shows the cluster profile. From the profile we can deduce that Group 1 has longer links than Group 2 because of its higher $\bar{y}$ value; also the links in Group 1 are more scattered in the y directions because of a higher standard deviation.

From the distribution we can see that Group 2 (35) contains more than double the links in Group 1 (17). The link index for Group 1 is 17/18 (0.94) and the link

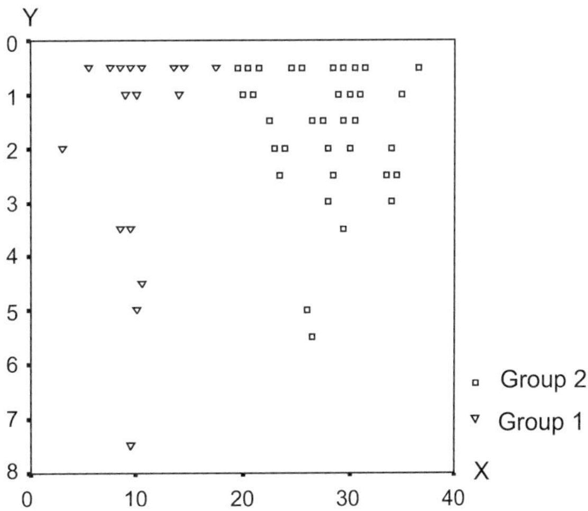

Fig. 3.16 Scatter plot of the two clusters generated by SPSS

Table 3.13 Cluster distribution of the linkograph

	N	% of combined	% of total
Cluster			
Group 1	17	32.7	32.7
Group 2	35	67.3	67.3
Combined	52	100.0	100.0
Total	52		100.0

Table 3.14 Cluster profile of the linkograph

Centroids	X		Y	
	Mean ($\bar{x}$)	Std. deviation	Mean ($\bar{y}$)	Std. deviation
Cluster				
Group 1	10.06	3.42	1.94	2.11
Group 2	27.81	4.59	1.67	1.24
Combined	22.01	9.41	1.76	1.56

Table 3.15 Hypothetical linkographs of five design moves and their interpretations

Case 1	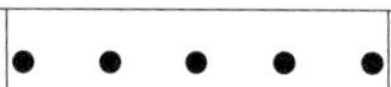	Five moves are totally unrelated, indicating no converging ideas, hence very low opportunity for idea development
Case 2	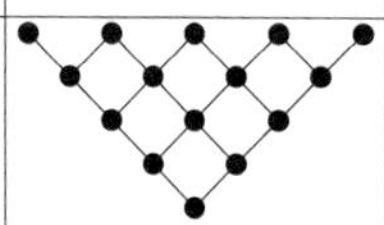	All moves are interconnected; this shows that this is a totally integrated process with no diversification, hinting that a premature crystallisation or fixation of one idea may have occurred, therefore there is a very low opportunity for novel ideas
Case 3	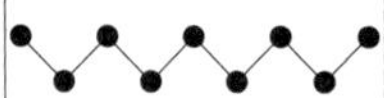	Moves are related only to directly preceding moves. This indicates the process is progressing but not developing, indicating some opportunities for idea development
Case 4	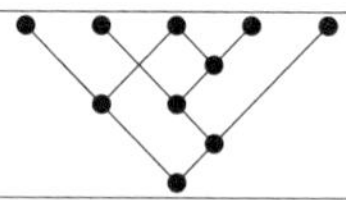	Moves are inter-related but not totally connected, indicating that there are lots of opportunities for good ideas with development

index for Group 2 is 35/19 (1.84). According to Goldschmidt (1992), the second half is more productive than the first half. The overall session link index is 52/37 (1.41). Essentially, the link index indicates the situation of a linkograph. From a theoretical viewpoint, is a saturated linkograph desirable? Does a fully linked linkograph indicate no diversification of ideas, hence less opportunity for creative outcome? This proposition is exaggerated with four hypothetical design scenarios in Table 3.15. We speculate that a partially linked linkograph embodies a balanced process in the sense that it embraces both integration and diversification of ideas. The figures in Table 3.15 suggest that the opportunity for idea development has some relationship with the predictability of the links in the linkograph. The links in Case 1 and Case 2 are predictable in the sense that they are either all linked or all unlinked. It is very easy to describe them. In Case 3 and Case 4 there are many more possibilities; more words are needed to describe them. The amount of words needed to communicate those linkographs directs the study to explore the use of the information theory of communication (Shannon 1948) to measure the graphs.

3.7 Information Theory

Shannon (1948), the founder of information theory, suggested that communication of information can be measured by the probability of its outcome and the semantics of information are irrelevant. The amount of information carried by a message or symbol is based on the probability of its occurrence. If the probability is 1, there is only one possible outcome, then there is no need to communicate additional information because the outcome is known. In the hypothetical cases in Table 3.15, there are ten possibilities of linkage. Cases 1 and 2 can be considered as all unlinked and all linked. Only one signal or symbol is needed to communicate them. In Cases 3 and 4 the probabilities of having a link are 4/10 and 5/10 respectively; more symbols are needed to communicate them. This section will propose how to use information theory to describe and measure a linkograph. It will start with the information-generation function and the calculation of entropy, which is the unit of measurement of information.

In Shannon's formulation of information theory, communication systems are modelled as a stochastic process (a simple definition of stochastic process is an ordered collection of random variables) of information transmitted from a source through a channel. Information is transmitted through recognisable symbols predetermined by the source and the receiver (encoding and decoding). If the outcome is known then there is no additional information. To illustrate this with a simple example, consider transmitting a piece of information consisting of ten ON/OFF signals and one of them is OFF but the others are ON. The probability of an OFF symbol, $p(\text{OFF})$, is 0.1 and the probability of an ON symbol, $p(\text{ON})$, is 0.9. Consider the following two cases:

1. If the first signal the receiver gets is an OFF symbol ($p = 0.1$), then no further transmission is required as the following signals carry no additional information. This, a stochastic process, assumes that the receiver knows the total number of signals (10), the probabilities of the symbols (ON/OFF), and that the total probability equals 1 ($p(\text{ON}) + p(\text{OFF}) = 1$).

2. If the first signal being transmitted is an ON symbol (p = 0.9), then the receiver is uncertain of the value of the next signal. Further transmission is still required to complete the information.

The transmission of the first case carries more information. The amount of information carried by a symbol (ON or OFF in this case) is related to the probability of its outcome.

Another example concerns the game of bridge. If a player calls something that surprises her/his partner, her/his partner gets more information. Based on these kinds of observations, Shannon proposed an information-generating function $h(p)$. This information function needs to have the following properties:

$h(p)$ is continuous for $0 \leq p \leq 1$, where p is the probability;
$h(p_i)$ = infinity if $p_i = 0$, where pi is the probability of a given state;
$h(p_i) = 0$ if $p_i = 1$;
$h(p_i) > h(p_j)$ if $p_j > p_i$, where p_i and p_j are the probabilities in two different states; and
$h(p_i) + h(p_j) = h(p_i \times p_j)$ if the two states are independent.

Shannon proved that the only function that satisfies the above five properties is:

$$h(p) = -\log(p) \tag{3.9}$$

Given a set of N independent states $a_1, \ldots a_n$ and the corresponding possibilities $p_1, \ldots p_n$, (in the above example, $N = 2$, $p_1 = p(\text{ON}) = 0.9$, and $p_2 = p(\text{OFF}) = 0.1$). Shannon derived entropy (H), the average information per symbol in a set of symbols, to be:

$$p_1 \times h(p_1) + p_2 \times h(p_2) + \cdots + p_n \times h(p_n) \tag{3.10}$$

Therefore

$$H = \sum_{i=1}^{n} p_i \{\log_b(p_i) \quad \text{with} \quad \sum_{i=1}^{n} p_i = 1 \tag{3.11}$$

In the example there are two symbols (ON/OFF) and entropy is expressed by:

$$H = -p(\text{ON})\log_2(p(\text{ON})) - p(\text{OFF})\log_2(p(\text{OFF})) \tag{3.12}$$

Substitute the values of probabilities:

$$H = -(0.9 \times \log_2(0.9) + 0.1 \times \log_2(0.1)) = 0.469 \tag{3.13}$$

The "logarithmic base corresponds to the choice of a unit for measuring information" (Shannon 1948, p. 379). Here base 2 is used to represent the binary (a binary system represents numeric values using two symbols, usually 0 and 1) ON and OFF information.

The next section describes how this can be applied to calculate the entropy of a linkograph of a design session.

3.7.1 Entropic Measurement of Linkographs

The authors consider an empty linked linkograph as a non-converging process with no coherent ideas and that a fully linked linkograph stands for a wholly integrated process with no diversification (refer to Table 3.12). In both cases the opportunities for idea development are very low. This line of reasoning can be expressed in terms of entropy; if a move is randomly picked from an empty linked linkograph, we can be certain that it is not linked to any other moves. This sounds obvious, but this linkograph can be considered as a carrier with zero information content; because the outcome is known, it will have zero entropy. Similarly, a fully linked linkograph will also have zero entropy.

In order for the entropy measurement of a linkograph to be meaningful, the conceptual differences between forelink and backlink must be considered. A third link type called a horizonlink is introduced. A horizonlink is not a link itself but it bears the notion of length of the links, which also maps onto time (separation) between links. It can be viewed as a measure of the distances of the links. It characterises two opposite notions: cohesiveness and incubation. Figure 3.17, where black dots denote linkages between moves and grey dots denote no linkage between moves, shows a typical linkograph with more cohesive links (short links) than incubated links (long links). When considering the short links, if ideas are not cohesive there is a lack of integrations hence they are not desirable. However, if ideas are too cohesive there is a lack of innovation. Similarly, totally connected long links indicate lack of diversification. In practice, however, long links are rare and are usually desirable, as they revisit previous ideas, which might indicate the importance of those ideas. Figure 3.18 shows three abstracted linkographs for entropy measurement;.

In Fig. 3.18c, it can be observed there are $n - 1$ rows in an n moves linkograph. Let $n - i$ denotes the row number; the links in rows with a small i indicate that the distance between moves is small, and they are labelled as short links. These moves will likely reside in working memory and are referred to as the cohesiveness of

Fig. 3.17 A linkography with typical distribution of links during a design process

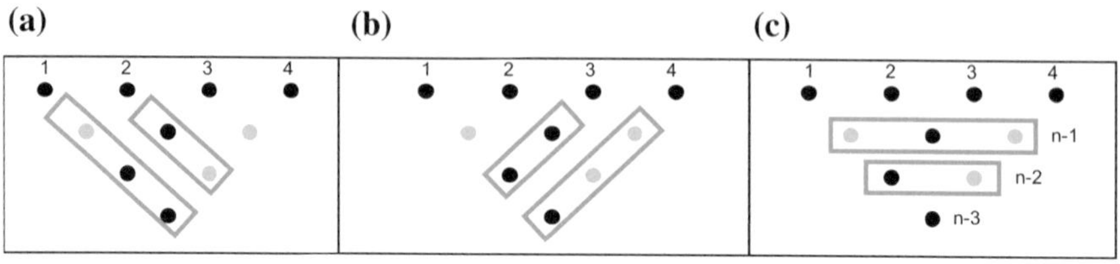

Fig. 3.18 Entropy measures of linkograph: **a** forelink, **b** backlink, and **c** horizonlink

ideas. However, if the ideas are too cohesive, that might imply fixation and lack of innovation. The links in the rows with a larger i connect moves that are far apart; they are called long links. These moves may not be in the working memory and are considered as incubated moves. Long links are comparatively rare and may signify reflection in action. The authors assume that a good design process is reflected in a linkograph that contains unsaturated short links (cohesive links) plus a number of long links (incubated links).

The forelink entropy for each move is computed by Eq. 3.12, except for the last two moves. The p(ON) represents the probability of linkage and p(OFF) represents the probability of no linkage. For the last two moves, as seen in Fig. 3.18a, move 4 will not have forelinks and move 3 is either linked or unlinked to move 4, which will have zero entropy. Similarly, each segment except the first two will receive a backlink entropy, Fig. 3.18b. The moves legitimate for entropy calculation are enclosed by rectangles in the figure. In move 1 there are three nodes for links inside the rectangle; move 1 and move 2 are unlinked, while move 1 is linked to move 3 and move 4. The percentage of linked nodes is 66.6 % and the percentage of unlinked nodes is 33.3 %. So the probability will be: $p(\mathrm{ON}) = 0.666$ and $p(\mathrm{OF\ F}) = 0.333$ respectively. If we substitute these in Eq. 3.12, the forelink entropy for move 1 becomes:

$$H = -0.666\log_2(0.666) - 0.333\log_2(0.33) = 0.918$$

Similarly, the forelink entropy for move 2:

$$H = -0.5\log_2(0.5) - 0.5\log_2(0.5) = 1$$

As for move 3, there is only one possible link. No matter whether it is ON or OFF, the probability is 1 and the entropy is zero, because $\log_2(1) = 0$.

Using this method, the backlink entropies for move 3 and move 4 in Fig. 3.18b are 0 and 0.918 respectively.

For the horizonlink entropy in this case, only two rows are considered: $n - 1$ and $n - 2$. If those are computed with Eq. 3.12, the entropy of the $n - 1$ row is 0.918 and the entropy of the $n - 2$ row is 1. Since people have limited short-term memory (Miller 1956), applying Miller's "magic number seven plus or minus two" objects, linkographs seldom have segments with more than nine links and the number of links between far-apart segments will decrease. Figure 3.14 shows a typical linkograph which has many cohesive links but very few incubated links. A fully cohesive link, for example, all ON in $n - 1$, will have 0 horizonlink entropy; similarly, if there are no incubated links, that row will score 0 in horizonlink entropy as well.

If an idea is not used, it will not have many forelinks and this is represented by low entropy. However, if an idea has too many forelinks, this might indicate fixation; this is also indicated by low entropy. Backlink entropy measures the opportunities according to enhancements or responses. If an idea is very novel, it will not have backlinks. The resulting entropy is low. On the other hand, if an idea is backlinked to all previous ideas, it is not novel. Hence, it is represented by low entropy. Horizonlink entropy measures the occurrence of incubated segments. Low horizonlink entropy indicates complete cohesiveness. Horizonlink entropy measures the opportunities relating to cohesiveness and incubation.

The proposition that an intensively linked linkograph indicates good designs should apply up to a certain point of saturation. In the early stages of designing, fixation is not desirable. Fixation is indicated by a move with near- saturated forelinks. Here the suggestion is that forelink entropy measures the idea-generation opportunities in terms of new creations or initiations. Figure 3.19 compares the link index measurement with the entropy measurement of a move based on Eq. 3.12. A heavily linked and a sparsely linked linkograph will have low entropy values. However, the link index increases as the number of links increases. The slope of the link index in Fig. 3.19 is not fixed, as it is determined by the total number of moves of the linkograph. In this particular graph, it can be observed that the link index matches, but not closely, the entropy until the graphs intersect at about 75 % of saturation. It is very rare to have linkographs over 10 moves with that level of saturation. After this point, entropy drops, while the link index value continues to increase.

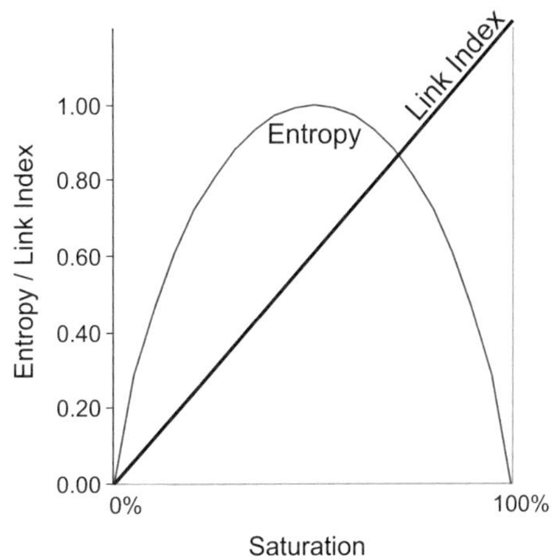

Fig. 3.19 Compare link index and entropy measurement

It can also be observed that the entropy curve in Fig. 3.19 is symmetrical; the slope of the graph decreases sharply as the probability moves away from 0 and 1. This indicates that when the links move away from determinate values of 0 and 1 (all un-linked and all linked), the H value increases rapidly. This graph shows that when $p(1)$ is between {0.35, 0.65}, H is over 0.93, that is, if the links in a row are between 35 and 65 %, it will produce a very positive value (rich design process). If the links are less than 5 % or over 95 %, it will produce a very low H value (below 0.29).

To illustrate the differences in these two measurements, link index and entropy, of linkographs, we use the hypothetical cases in Table 3.15 again. In Case 1 the probability of ON for all moves is 0 and the probability of OFF is 1, put these in Eq. 3.12, H = 0 because $\log_2(1) = 0$. Therefore the entropies will be 0 for any moves in any direction, hence the cumulative entropies will be 0.

For Case 2 the probability of ON for all moves is 1 but the probability of OFF is 0, so again, similar to Case 1, the cumulative entropies will be 0.

In Case 3, consider the forelink entropy of:

- the first move, there is one link out of four possible links, therefore the p(ON) = 1/4 = 0.25 and p(OFF) = 3/4 = 0.75, so H = 0.81;
- the second move, there is one link out of three possible links, p(ON) = 1/3 = 0.33 and p(OFF) = 2/3 = 0.67, so H = 0.92;
- the third move, there is one link out of two possible links, p(ON) = 0.50 and p(OFF) = 0.50, so H = 1.00;
- the fourth move only has one possible link, so no matter it is ON or OFF entropy value will be zero;
- the fifth move does not have any forelinks, so no entropy value.

The cumulative forelink entropy will be 0.81 + 0.92 + 1 = 2.73. As for the backlink cumulative entropy, the calculation will be similar to that of the cumulative forelink entropy but in the reverse order and with the same values.

The horizonlink entropy will be calculate by rows; starting from the bottom, the first row has only one possibility of ON and OFF so the entropy is zero. The second row has two possible links and both are OFF, so the entropy is zero again; this is the same for the third row. As of the fourth row, there are four possible ON and OFF links, in the case all are ON and the entropy is zero. Therefore the cumulative horizonlink entropy is zero.

In Case 4, consider the forelink entropy of:

- the first move, there are two links out of four possible links, therefore the p(ON) = 2/4 = 0.50 and p(OFF) = 2/4 = 0.50, so H = 1.00;
- the second move, there are one links out of three possible links, p(ON) = 2/3 = 0.67 and p(OFF) = 1/3 = 0.33, so H = 0.92;
- the third move, there is one link out of two possible links, p(ON) = 0.50 and p(OFF) = 0.50, so H = 1.00;
- the fourth and fifth moves have no entropy value.

The cumulative forelink entropy will be 1.00 + 0.92 + 1.00 = 2.92.

Consider the backlink entropy of:

- the first move, does not have any backlink, so no entropy value;
- the second move has only one possibility for link, so entropy will be zero;
- the third move, there is one link out of two possible links, so p(ON) = 0.50, p(OFF) = 0.50 and H = 1.00;
- the forth move, there are two links out of three possible links, p(ON) = 2/3 = 0.67 and p(OFF) = 1/3 = 0.33, so H = 0.92;
- the fifth move, there are one links out of four possible links, therefore the p(ON) = 2/4 = 0.50 and p(OFF) = 2/4 = 0.50, so H = 1.00.

Therefore the cumulative backlink entropy for Case 4 is 1.00 + 0.92 + 1.00 = 2.92.

Consider the horizonlink entropy of:

- the bottom row, which has only one possible links, therefore no matter it is linked or not linked the entropy will be zero;
- the second row, there is one link out of the two possible links, so p(ON) = 0.50, p(OFF) = 0.50 and H = 1.00;
- the third row, there are two links out of three possible links, p(ON) = 2/3 = 0.67 and p(OFF) = 1/3 = 0.33, so H = 0.92;
- the fourth row, there is one link out of four possible links, therefore the p(ON) = 1/4 = 0.25 and p(OFF) = 3/4 = 0.75, so H = 0.81.

Therefore the cumulative horizonlink entropy for Case 4 is 1.00 + 0.92 + 0.81 = 2.73.

The link index of Case 1 equal zero because there is no links. There are 10 links in Case 2 so the link index is 2 (10/5, i.e. 4 links divided by five moves).

The link index of Case 3 equals 4/5 (4 links divided by 5 moves) and the link index of Case 4 is 5/5 = 1.

Table 3.16 compares the link index values and the entropy values of the hypothetical case depicted in Table 3.15. The entropy values in the table are the

Table 3.16 Comparison of the cumulative entropy and link index of hypothetical case

	Case 1	Case 2	Case 3	Case 4
		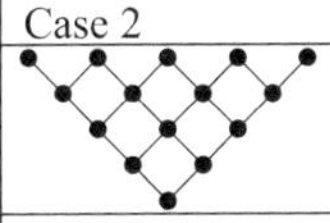		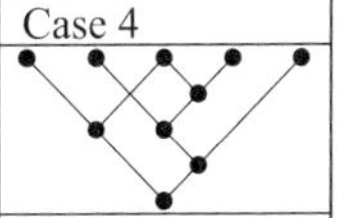
Forelink H	0.00	0.00	2.73	2.92
Backlink H	0.00	0.00	2.73	2.92
Horizonlink H	0.00	0.00	0.00	2.73
Total H	0.00	0.00	5.46	8.55
Link index	0.00	2.00	0.8 0	1.00

cumulative values of the contribution of each moves. The total value is the addition of the three different types of entropy. Link index benchmark Case 2 is the most productive scenario. However, as explained in Table 3.15, this might not be the most desirable scenario.

3.7.2 Normalizing Entropic Measurement for Comparison

Table 3.16 shows the cumulative values of entropy, which will increase as the number of moves increase. It is possible to calculate the maximum entropy of a linkograph and normalize against it by dividing the entropy calculated by the maximum entropy. This will always give us a number less than or equal to one. In our calculation, the maximum entropy occurs when the probability of link and unlink have the same value. i.e. when $p(\mathrm{ON}) = p(\mathrm{OFF}) = 0.5$, $H = 1$. This happens when there are even numbers of possible links, as we can observe only about half of the graph will have this maximum entropy of 1 and the remaining moves will always have an entropy of less than 1. So for a graph with n moves the maximum forelink and backlink entropy are $(n - 1)$ and the maximum horizonlink is $(n - 2)$. Table 3.17 shows the normalized entropy of the example in Table 3.15.

In summary, this section proposes using entropy to measure and study linkographs, in addition to link index and critical move analysis. Also, it describes how to calculate the entropy of a linkograph. The contribution of each move is counted in three different ways: according to forelinks, backlinks and horizonlinks. It is hypothesized that entropy measures the idea development opportunities. Forelink entropy measures the idea-generation opportunities in terms of new creations or initiations. Backlink entropy measures the opportunities according to enhancements or responses. Horizonlink entropy measures the opportunities relating to cohesiveness and incubation. Further, it is hypothesised that the entropy measurement of a linkograph is positively correlated to the design outcome, due to better opportunities for idea development.

Table 3.17 Normalized entropy of hypothetical case

	Case 1	Case 2	Case 3	Case 4
	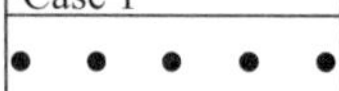	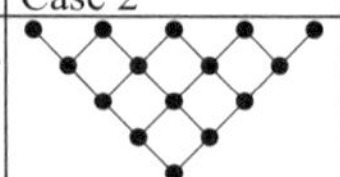	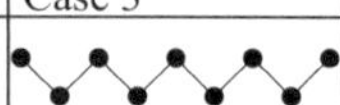	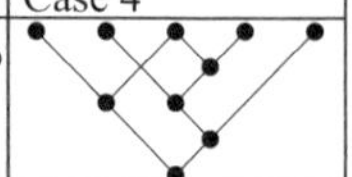
Forelink H	0.00	0.00	0.68	0.73
Backlink H	0.00	0.00	0.68	0.73
Horizonlink H	0.00	0.00	0.00	0.91

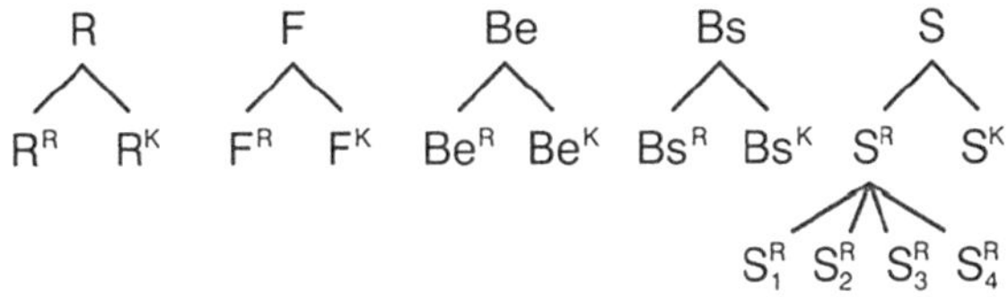

Fig. 3.20 Expanding FBS scheme to cover subclass variables (after Yu et al. 2013)

3.8 Summary

This chapter has proposed an ontological coding scheme and described its application. As an example of how this coding scheme can be applied more widely, Yu et al. (2013) utilised the FBS coding scheme to study designer's behaviour in a parametric design environment by instantiating the codes into multiple subclasses, Fig. 3.20. This does not modify the ontology as the subclasses with their own codes can be aggregated back into the FBS primary codes. They defined two types of design spaces: design knowledge space (denoted by the superscript K) and rule algorithm space (denoted by the superscript R). In the design knowledge space designers make use of their design knowledge and in the rule algorithm space designers apply design knowledge through the operations of parametric design tools. The structure variables in the rule algorithm can have more subclasses of the specific rule algorithm activities in the parametric design environment. By doing this, distinct activities can be mapped back to the FBS class variable and comparisons can be made with other design situations.

This chapter also revisited the linkography technique. Syntactic and semantic design processes have been defined. Statistical analysis methods have been depicted as a means to a describe design session and to produce inference-based observations. Markov chains have been proposed to study the design protocol as a time series, which provides another venue for examining design protocol data. In addition, two mathematically based methods, statistical clustering and Shannon's entropy, were proposed to analyse linkographs. This chapter has covered the theoretical background of using these quantitative methods in addition to those traditional methods of linkography to investigate design protocols.

Chapter 4
Ontologically-Based Studies of Design Protocols

This chapter presents the results of using an ontological approach to segment and code the protocol before constructing a linkograph. Design processes are then derived from the links. Four cases are being presented. The first case was a brainstorming session that involved a multi-disciplinary design team. In this case, the percentage of processes derived from the FBS coding scheme is compared with those derived from the situated FBS coding scheme. In the second case two architects collaborated in two different environments; one a face-to-face session and the other using a 3D virtual-world environment; the analyses of the two sessions are performed by using the situated FBS ontology. The third case uses statistical inferences to explore the effects of education on design. In case four, designing is viewed as a special class of problem solving; the FBS design issues is being mapped onto problem and solution spaces. This forms the basis of using the metacognitive design style of students to compare designerly behaviour in different domains.

4.1 Case One: A Brainstorming Session

The data for this case was an in situ design meeting distributed to researchers involved in the Design Thinking Research Symposium 7. The idea behind using a common set of protocol data was to find a more rigorous way to do empirical research into design (Cross 2007b). There were four sessions, two architectural and two engineering. One particular engineering session was selected because of its content. Since the architectural sessions contained mainly presentation and communication with clients, they contained comparatively fewer design activities. Both engineering sessions concerned brainstorming of a thermal printing pen, the first session was selected because it involved generating novel ideas related to the structure of an object. The other session focussed more on usage and control.

J.W.T. Kan and J.S. Gero, *Quantitative Methods for Studying Design Protocols*, DOI 10.1007/978-94-024-0984-0_4

Fig. 4.1 Four camera digital recording of the brain storming session

The aim of this brainstorming session was to obtain ideas for a prototype of the thermal printing pen. This involved solving specific problems such as keeping the print head levelled with an optimum angle and protection of the print head. Seven cross-disciplinary participants were involved, with one acting as the moderator. The whole session lasted for 1 h and 37 min. Figure 4.1 shows a frame from the recording.

4.1.1 Qualitative Analysis

This protocol can be divided into two episodes; the first one concerned the problem of keeping the print head in contact with and at the optimum angle to the media, despite wobbly arm moment. The second episode dealt with protecting the print head from abusive use and overheating. In the first episode, participants were asked to generate ideas from available products that follow a contour. Several products were mentioned, such as a sledge, snowboard, wind surfboard, shaver, snow-mobile, train and slicer. Other concepts such as wheels, spirit level and laser leveller were also discussed. Loosely related to those analogies, a few proposed shapes, such as a mouse-type pen, were proposed. Besides product behaviour, user behaviour was also considered. Figure 4.2 shows some of the sketches the

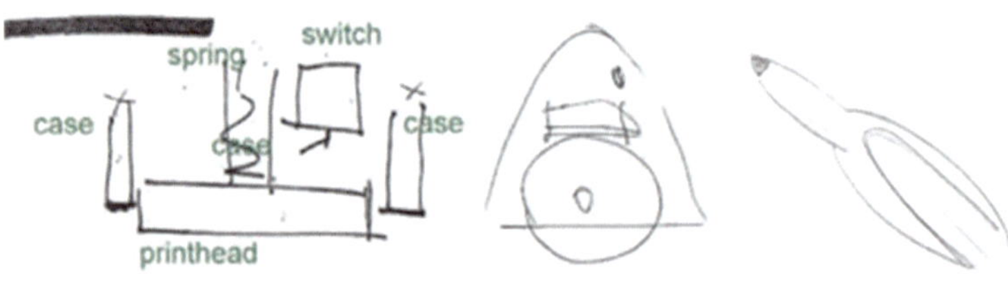

Fig. 4.2 Sketches from the brainstorming session

participants used in this session. On the left is a sketch of the structure they proposed near the end of the first episode; in the middle is a drawing of a toy a member suggested from which to borrow ideas; and on the right is one of the proposed forms of the thermal printing pen.

4.1.2 Segments and Coding

The verbal transcription of the protocol was segmented strictly according to the FBS ontology—each segment contains only one FBS code that represents the design issue in that segment. Segmenting and coding were undertaken simultaneously by discerning whether an action or utterance expresses the FBS aspect of designing. If an utterance contains more than one FBS issue it will be further divided. In the first round, the protocol was coded with the original FBS classes. Two additional codes were used: requirement (R) and others (O) to represent issues related to requirements and any other non-design related issues. The addition of the R code does not require any addition to the FBS ontology as R is expressible in terms of F, B or S. In Gero's (1990) FBS computational model, designing was assumed to start with function. Later in the Gero and Kannengiesser's (2004) situated FBS framework (a cognitive model), designing was viewed to start with requirements. The R code is used because in protocol studies the designing activities usually start with requirements instead of function. Table 4.1 shows some examples from the protocol for each code and Fig. 4.4 contains an extract of the coded protocol used to show the linkograph.

The first episode was coded twice by the first author with a 10-day separation and then self-arbitrated using the Delphi method proposed by Gero and McNeill (1998). The agreement between the codes is 86 %. Within the 52 min in this

Table 4.1 Example of coding

Issues	Protocol
Requirements (R)	"quite important is it's about the thermal-incli-inclis () pen" "design a-a prototype"
Function (F)	"that's the standard plain thermal paper err and then it can draw"
Expected behaviour (Be)	"either atoms or line types" "we can print thermo reactive dyes onto media substrates"
Behaviour (Bs)	"it'll be about 50 % more expensive" "if you lift an optical mouse slightly off the page you'll see the pattern it creates"
Structure (S)	"a sledge or a snowboar- a skis or snowboard" "show the relative size of the pen if you've got an example"
Design description (D)	Figure 4.2
Others (O)	"yeah, we'll come to that in a minute"

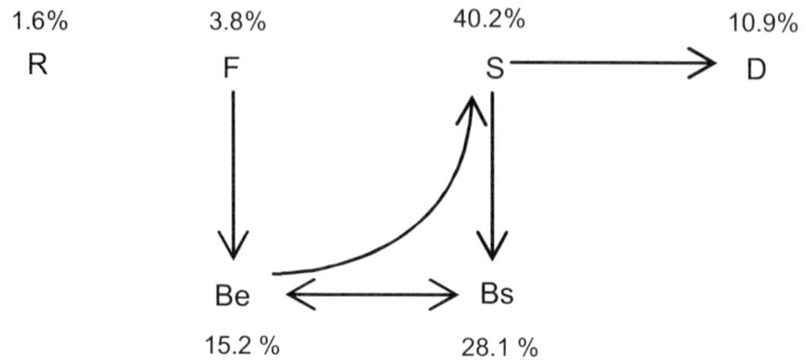

Fig. 4.3 Percentages of each code in relation the FBS framework

episode there were 475 segments. The average segment length is about 6.5 s. Of the segments, 448 segments have FBS issues; those segments without FBS issues (27) consist mostly of jokes or communications that are not related to the design process or the resulting artefact. Figure 4.3 shows the percentages in each of the FBS categories in relation to the processes of the FBS ontology. The highest percentages are in the structure and behaviour classes. In this protocol these high percentages were due to the frequent use of analogies with other products and situations.

4.1.3 Linking the Segments

In the first run of constructing the links, the connections between the segments were discerned independently of the code. After 4 months, the links were re-examined in conjunction with the codes. This was to increase the reliability in linking the segments. Figure 4.4 presents an extract of the coded protocol together with its linkograph. In the table part of Fig. 4.4 column one is the segment number, column two is the code and column three is the transcribed protocol.

In this extract two participants were involved, the moderator (A) and a mechanical engineer (J). The focus of the discussion was "other products or situations where a product needs to follow a contour". J suggested an object (structure) —"sledge" (segment 38)—and continued to explain the behaviour of the sledge, i.e., how it maintains contact or level on the snow (segment 40 and 48). The sledge was compared with a set of skis (segment 43) in terms of the structure (segment 44) and behaviour (segments 45, 47 and 48). The coding of segment 50 can be argued; it was coded as expected behaviour (Be) as we interpreted J was borrowing the behaviour of the analogised objects and targeting those to be the expected behaviour of the designed object. Finally, the structure of stabilisers (segment 53) was suggested. Segment 39 was linked to segment 38 because the "write sledge" action was a response to the initiation and suggestion of the "sledge" in segment 38.

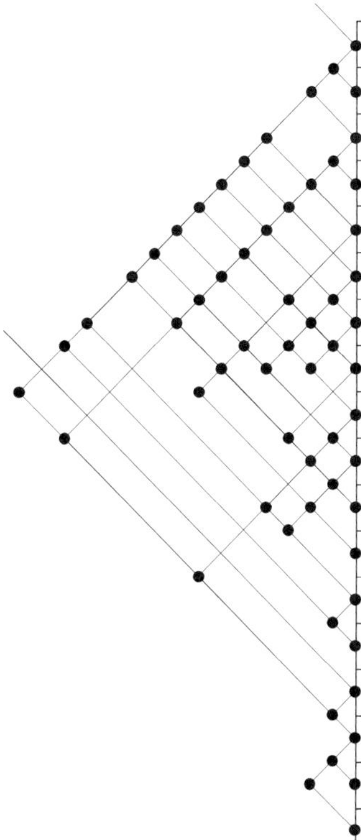

38	S	J: I ended up with the + hold on +sledge
39	D	A: the sledge excellent so what did that generate then? (write: sledge)
40	Bs	J: the sledge manages to keep level by having quite a wide base
41	D	A: (write: wide base)
42	Bs	J: a main force in the middle
43	S	J: unlike the set of skis
44	S	J: where quite narrow and
45	Bs	J: you go up on an edge when you're turning
46	S	J: the sledge is er quite broad
47	Bs	J: and then you have the weight right in the middle
48	Bs	so they manage to keep both runners on the snow
49	D	A: (write: force in middle)
50	Be	J: a sledge or a snowboar- a skis or snowboard
51	S	A: some guiders almost down the side of this
52	Be	J: the easiest way to keep the pen at a right angle would be
53	S	J: to have a set of stabilisers on it based on the idea of a sledge
54	S	A: stabilisers +++like a bicycle yeah that's a good
55	D	A: (write: stabiliser)

Fig. 4.4 Rotated linkograph in relation to the protocols

J started explaining in segment 40 why a sledge was a proposed candidate for solution so segments 38 and 40 were linked. By examining the relationship of a segment with those preceding segments a linkograph was constructed. Figure 4.5 shows a larger part of the linkograph of this session that includes the above extract. Other clusters were also labelled. These clusters can be distinguished by visual inspection of link density or by statistical clustering.

There are 2110 links connecting the 448 segments, so on average each segment has 4.7 links. However, some segments have many more links than others. Table 4.2 compares the distribution of the codes of the segments with the occurrences of codes in the links. It shows the percentages of segments and links with the FBS code. Compared to the coded segments, it can be observed that the codes in the links have a decrease in the documentation, a moderate decrease in behaviour derived from structure and a slight decrease in function. The requirement has increased, and there is an increase in expected behaviour and structure as well. This implies that the expected requirements, behaviour and structure segments in this session are more influential because the designers took more notice of those linked segments.

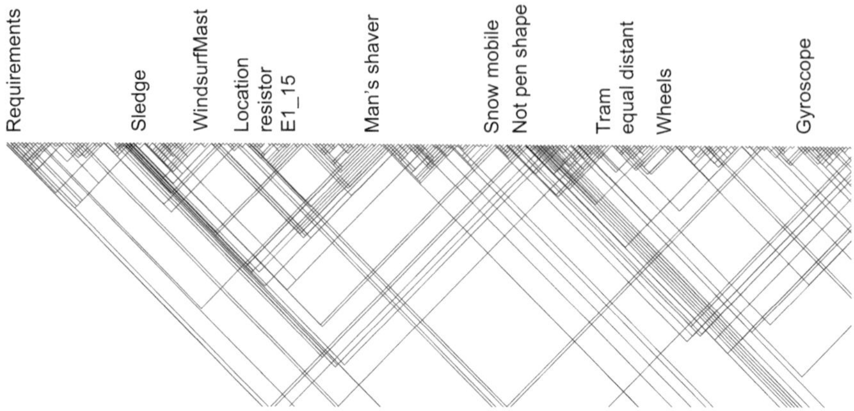

Fig. 4.5 Part of the linkograph of the segmented protocol

Table 4.2 The distribution of codes in segments and links

Code	Segment		Link	
	Count	Percentage	Count	Percentage
R	7	1.6	36	1.7
F	17	3.8	56	2.6
Bs	126	28.1	504	23.8
Be	69	15.4	396	18.7
S	180	40.2	936	44.3
D	49	10.9	187	8.8
Total	448	100.0	2110	100.0

4.1.4 *Deriving Semantic FBS Processes from Coded Segments and Links*

In the following analysis, the symbol "⇝" is used to denote the link between the nth and the (n + i)th segments to avoid confusion with the transformation symbol "→" (see Chap. 3, Sect. 3.1.2). For example, consider the first segment linking to the two subsequence segments in Fig. 4.4, S ⇝ D was used to represent the link between segments 38 and 39. S ⇝ Bs was used to represent the link between segments 38 and 40, Fig. 4.6 illustrates this example. S ⇝ D can be seen as the documentation process (transformation from structure to design description, S → D) and the S ⇝ Bs as the analysis process (transformation structure to behaviour, S → Bs) according to the ontology. There may be cases where the links might not correspond to the eight FBS processes.

The 2110 links can be viewed as design processes since each link has an FBS code at each end. Thus, looking back from one coded segment to the segment at the other end of the link we have a transformation from one coded issue to another, i.e.,

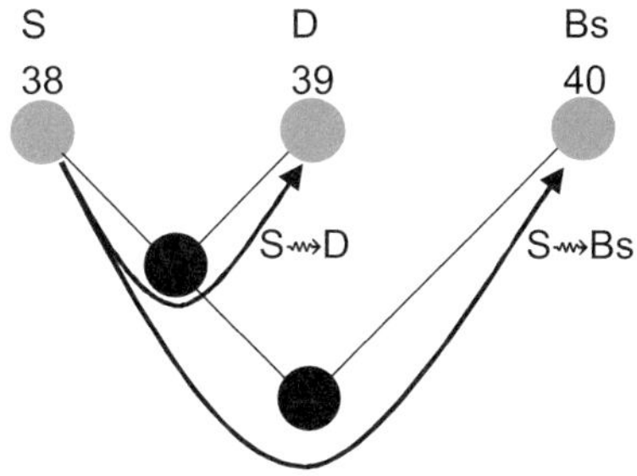

Fig. 4.6 Deriving transformation processes from linkograph

a design process, and the linkograph then becomes a network of transformation processes. There are seven categories of codes, including O, so there will be 49 types of possible transformations. However, according to the FBS ontology, many of those processes have no direct meaning in design. For example, a process that involves O will have no design significance. Table 4.3 shows the FBS-related processes derived from the links of the segments with FBS issues. There are 30 types of FBS processes recorded; those FBS processes in the framework are represented in Table 4.4.

In this episode, the reformulations were mostly of structure and behaviour. The sledge example in Fig. 4.4 contains the reformulation of structure (S ⇝ S), from the structure of a sledge to the structure of a set of stabilizers like those in a bicycle, segment 38–53 and 53–54.

Table 4.3 Percentages of all the processes derived from codes and links

R		F		Bs		Be		S		D	
R ⇝ Bs	1.3	F ⇝ Bs	0.2	Bs ⇝ Bs	8.4	Be ⇝ Bs	4.9	S ⇝ Bs	12.2	D ⇝ Bs	0.5
R ⇝ Be	0.2	F ⇝ Be	1.1	Bs ⇝ Be	5.0	Be ⇝ Be	6.9	S ⇝ Be	3.3	D ⇝ Be	0.9
R ⇝ F	0.1	F ⇝ F	1.4	Bs ⇝ D	0.9	Be ⇝ D	1.8	S ⇝ D	5.3	D ⇝ D	1.4
R ⇝ R	0.3	F ⇝ R	0.1	Bs ⇝ F	0.5	Be ⇝ F	0.1	S ⇝ F	0.1	D ⇝ S	5.4
R ⇝ S	1.0	F ⇝ S	0.1	Bs ⇝ S	2.7	Be ⇝ S	6.9	S ⇝ R	0.1		
								S ⇝ S	26.7		

Table 4.4 Percentages of the eight FBS processes

Processes		Occurrence	Percentage
Formulation	R → F, F → Be	14	1.4
Synthesis	Be → S	68	6.9
Analysis	S → Bs	120	12.2
Documentation	S → D	52	5.3
Evaluation	Be ↔ Bs	48(Be ⇝ Bs), 49(Bs ⇝ Be) 97	9.9
Reformulation I	S → S	262	26.7
Reformulation II	S → Be	32	3.3
Reformulation III	S → F	1	0.1
	Total	646	65.9

Other examples of structure reformulations were: making analogies with other products, for example a wind surfboard mast and man's shaver and considering the thermal pen in the shape of other things instead of a pen. Examples of behaviour reformulations were: using a universal joint to keep the angle; using springs to keep it level; and suggesting the locations of resistors, which prompted the responses regarding the cost.

The reformulation of function was rare, which reflects the nature of this session —mechanical brainstorming for ideas to keep the thermal pen in contact with the media at the correct angle. Some of the function aspects are deliberately not dealt with. For example the suggestion of "could we, sorry, could we actually see what they're doing? I mean, are they drawing pictures or making invitations or Christmas cards or—" was given the response of "erm we're going to try to deal with that a fair bit on Monday".

The FBS ontology covers two-thirds of the processes derived from the links of the coded segments. The segments that are not design related can be coded as "Other" with an "O". These segments can be deleted from the segmented/coded protocol when carrying out only design-related analyses. If we delete segments coded as "O" the coverage increases to three-quarters. Some of the most frequent processes not counted were: Bs $\rightsquigarrow$ Bs (8.4 %), Be $\rightsquigarrow$ Be (6.9 %), D $\rightsquigarrow$ S (5.4 %), S $\rightsquigarrow$ Be (3.3 %), Bs $\rightsquigarrow$ S (2.7 %), and Be $\rightsquigarrow$ D (1.8 %).

Reviewing the protocol, in the case of Bs $\rightsquigarrow$ S, it can be noted that the large scale of the granularity fails to pick up the Be in the Bs $\rightsquigarrow$ S processes. If the granularity of segmentation were finer, there should be an expected behaviour before the structure code. Using the example in Fig. 4.4, segment 40: "the sledge manages to keep level by having quite a wide base" was coded as Bs because it analyzes an existing product to get the "keep level" behaviour. This segment was linked to segment 53: "the easiest way to keep the pen at a right angle would be to have a set of stabilizers on it based on the idea of a sledge" which was coded as structure because it proposed a structure, "a set of stabilizers". The idea of sledge, the behaviour of "keep level", was translated to expected behaviour of "at a right angle" which leads to the structure of "a set of stabilizers".

The F $\rightsquigarrow$ F and Be $\rightsquigarrow$ Be can be viewed as reflections of function and behaviour in many cases. An example of Be $\rightsquigarrow$ Be happened when they discussed that the shape of the designed object does not need to resemble a pen. The moderator suggested that "…something else that gets pulled behind it for example" and an engineer's response, "…what they'll do is move the lump around" were linked and both segments were coded as expected behaviour (Be).

The D $\rightsquigarrow$ S is the interpretation of depicted structure. The Bs $\rightsquigarrow$ Bs usually is a result of further analysis, for example in Fig. 4.4 link between segments 40 and 42 and link between segments 42 and 48 were further analysis of the action and reaction of the force (weight). Sometimes the Be $\rightsquigarrow$ D transformation was the depiction of behaviour but the FBS ontology does not distinguish depiction of behaviour from depiction of structure unless we use subclasses as shown in Fig. 3.20. Using the above example again, "…what they'll do is move the lump around" was linked to the following segment where the moderator was writing down "(move lump)", which is a depiction of behaviour. These transformations are meaningful processes resulting from the interactions among members and artifacts. The situated FBS ontology covers all these processes.

4.1.5 Syntactic Design Processes: Markov Analysis

There are 475 segments and 448 of them contain situated FBS codes. If they are considered as first-order Markov processes, there will be 447 state changes. Table 4.5 lists the occurrence of all those state changes. There is no record of any Fe occurrence, so it was taken out for ease of matrix manipulation. Using the data in Table 4.5, the transition matrix, Eq. 4.1, can be obtained. The highest probability of transition was from B^e to Be^i. This indicates that an event of depiction of behaviour has a near 50 % chance to be followed by an event of expected behavior as shown in the results in Eq. 4.1.

$$P_{dtrs} = \begin{pmatrix} & R & F^i & Fe^i & B^i & Be^i & B^e & S^i & Se^i & S^e \\ R & 0.43 & 0.00 & 0.14 & 0.29 & 0.14 & 0.00 & 0.00 & 0.00 & 0.00 \\ F^i & 0.00 & 0.13 & 0.25 & 0.25 & 0.00 & 0.00 & 0.38 & 0.00 & 0.00 \\ Fe^i & 0.00 & 0.44 & 0.00 & 0.00 & 0.44 & 0.00 & 0.00 & 0.11 & 0.00 \\ B^i & 0.00 & 0.02 & 0.03 & 0.38 & 0.19 & 0.02 & 0.25 & 0.08 & 0.03 \\ Be^i & 0.01 & 0.00 & 0.01 & 0.19 & 0.28 & 0.12 & 0.04 & 0.35 & 0.00 \\ B^e & 0.00 & 0.00 & 0.08 & 0.15 & 0.46 & 0.00 & 0.00 & 0.31 & 0.00 \\ S^i & 0.02 & 0.00 & 0.00 & 0.39 & 0.03 & 0.00 & 0.32 & 0.17 & 0.07 \\ Se^i & 0.00 & 0.00 & 0.00 & 0.17 & 0.12 & 0.04 & 0.19 & 0.20 & 0.28 \\ S^e & 0.00 & 0.00 & 0.00 & 0.20 & 0.06 & 0.00 & 0.40 & 0.29 & 0.06 \end{pmatrix} \tag{4.1}$$

Table 4.5 Occurrence of the sequence of situated FBS codes

State	Next state								
	R	F^i	Fe^i	B^i	Be^i	B^e	S^i	Se^i	S^e
R	3	0	1	2	1	0	0	0	0
F^i	0	1	2	2	0	0	3	0	0
Fe^i	0	4	0	0	4	0	0	1	0
B^i	0	3	4	47	24	2	31	10	4
Be^i	1	0	1	13	19	8	3	24	0
B^e	0	0	1	2	6	0	0	4	0
S^i	2	0	0	38	3	0	31	17	7
Se^i	0	0	0	14	10	3	16	17	23
S^e	0	0	0	7	2	0	14	10	2

Probability Vector

Putting P_{dtrs} into Eq. 3.2 and solving the equations, the probability vector is:

$$a_{dtrs} = \begin{pmatrix} R & F^I & Fe^I & B^I & Be^I & B^e & S^I & Se^I & S^e \\ 0.01 & 0.02 & 0.02 & 0.28 & 0.15 & 0.03 & 0.22 & 0.19 & 0.08 \end{pmatrix} \quad (4.2)$$

This distribution is similar to the distribution of codes in segments. With a large number of segments, Eq. 4.2 predicts that 28 % of the codes are B^i, 22 % of the codes are S^i, 19 % of the codes are Se^i, 15 % of the codes are Be^i, 8 % of the codes are S^e, 3 % of the codes are B^e, 2 % of the codes are F^i, 2 % of the codes are Fe^i and 1 % of the code is R.

First Passage Times

Using P_{dtrs} and α_{dtrs} together with Eqs. 3.4 and 3.5, the mean first passage times can be obtained:

$$M_{dtrs} = \begin{pmatrix} & R & F^i & Fe^i & B^i & Be^i & B^e & S^i & Se^i & S^e \\ R & 84.7 & 66.1 & 43.7 & 4.3 & 6.9 & 35.8 & 7.6 & 7.5 & 15.4 \\ F^i & 146.8 & 56.8 & 41.7 & 4.1 & 8.8 & 36.8 & 4.5 & 7.3 & 14.6 \\ Fe^i & 147.2 & 42.0 & 51.3 & 5.4 & 5.8 & 34.9 & 6.5 & 6.2 & 14.3 \\ B^i & 146.9 & 70.9 & 55.5 & 3.6 & 7.5 & 35.1 & 5.2 & 6.3 & 13.5 \\ Be^i & 145.3 & 73.7 & 56.8 & 4.7 & 6.5 & 30.8 & 6.6 & 4.3 & 13.0 \\ B^e & 147.2 & 72.1 & 53.7 & 4.9 & 5.2 & 34.1 & 6.9 & 4.4 & 13.2 \\ S^i & 143.9 & 73.6 & 57.7 & 3.3 & 9.0 & 36.5 & 4.6 & 6.0 & 12.6 \\ Se^i & 147.1 & 74.5 & 58.5 & 4.4 & 8.4 & 34.9 & 5.1 & 5.3 & 9.8 \\ S^e & 146.6 & 74.4 & 58.5 & 4.1 & 9.0 & 36.5 & 4.1 & 5.2 & 12.3 \end{pmatrix} \quad (4.3)$$

The shortest mean passage time is from S^i to B^i and the longest one is from Fe^i to R. This matrix agrees with the general hypothesis that it is faster to move from any behaviour states to a structure state than from any function states to a structure state.

These statistically-derived models of designing provide a detailed level of understanding based on empirical evidence rather than relying only on a qualitative assessment. As a consequence it becomes possible to compare quantitative results from disparate studies (Gero and Jiang 2016).

4.1.6 Deriving Situated FBS Processes

The segments were recoded with the situated FBS scheme; readers can refer to Chap. 3, Sect. 3.4.3, for the ten categories. For ease of comparison, the segments have not been refined, so the total numbers of segments and total number of links

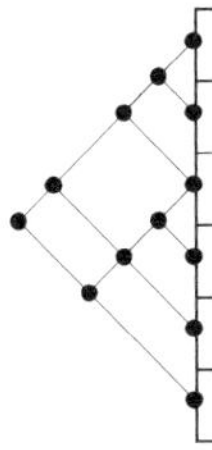

38	S^i	J: I ended up with the + hold on +sledge
39	S^e	A: the sledge excellent (write: sledge) so what did that generate then?
40	B^i	J: the sledge manages to keep level by having quite a wide base
41	S^e	A: (write: wide base)
42	B^i	J: a main force in the middle
43	S^i	J: unlike the set of skis

Fig. 4.7 Rotated linkograph showing the situated FBS coding of the protocol

will be the same. The protocol was coded twice by the first author and then self-arbitrated.

Figure 4.7 illustrates the recoding from segments 38–43. Segment 38 was coded as S^i because J was showing a picture of a sledge and was about to draw an analogy with the structure of the sledge. The main activity in segment 39 was writing down the word "sledge". It was treated as the documentation of structure as the word "sledge" denoted the object, so it was coded as S^e. Segment 40 was coded as Bi because it interpreted behaviour ("keep level") of the sledge. Segment 41 was coded as Se since it was a depiction that concerns the structural aspect ("wide base") of the object. Segment 42, similar to segment 40, involves the interpretation of behavioural aspect ("a main force in the middle") of the object, so the Bi code was assigned. Segment 43 concerns another object "skis" which was coded as S^i.

Table 4.6 shows the distributions of codes in the segments and links and their percentages. There is no documentation of function. It shows that expected behaviour, expected structure and interpreted structure are more influential than they appear in the segments, while the interpreted function and behaviour, the expected

Table 4.6 Comparing the distribution of codes in segments and links

	Segments		Links	
	Count	%	Count	%
R	7	1.6	36	1.7
F^i	8	1.8	18	0.8
F^e	0	0.0	0	0.0
Fe^i	9	2.0	38	1.8
B^i	125	27.9	493	23.3
B^e	13	2.9	42	2.0
Be^i	69	15.4	396	18.7
S^i	98	21.9	485	22.9
S^e	36	8.0	145	6.9
Se^i	83	18.5	462	21.8
Total	448	100.0	2126	100.0

Table 4.7 Percentages of FBS processes with situated FBS coding

Process		%
Formulation	R ⇝ B^i, R ⇝ F^i, R ⇝ S^i, Fe^i ⇝ Be^i, F^i ⇝ Be^i	3.4
Synthesis	Be^i ⇝ Se^i, B^i ⇝ Se^i	7.5
Analysis	S^i ⇝ B^i, S^e ⇝ B^i, Se^i ⇝ B^i	13.4
Evaluation	Be^i ⇝ B^i	4.6
Documentation	Be^i ⇝ B^e, B^i ⇝ B^e, Be^i ⇝ S^e, Se^i ⇝ S^e, S^i ⇝ S^e	8.1
Reformulation I	Se^i ⇝ Se^i, Se^i ⇝ S^i, S^i ⇝ Se^i, S^i ⇝ S^i, S^e ⇝ S^i, S^e ⇝ Se^i	31.7
Reformulation II	Be^i ⇝ Be^i, B^i ⇝ Be^i, B^i ⇝ B^i, B^e ⇝ Be^i B^e ⇝ B^i	21.6
Reformulation III	B^i ⇝ F^i, Fe^i ⇝ Fe^i, Fe^i ⇝ F^i, F^i ⇝ Fe^i	1.5
Total		91.9

function, and the depiction of behaviour are of less importance than they appear in the segments.

Using the situated FBS variables as the codes, 50 types of derived FBS-related processes were recorded. Those meaningful processes were then aggregated into the basic eight design processes. For clarity and ease of analysis, the interpretation and reflection categories of processes were separated from the formulation and reformulation processes, so that there is no overlapping of processes in any of the categories. This gave a 92 % coverage of the derived processes, as seen in Table 4.7. The middle column of the table represents the derived processes that were being aggregated. The percentages of all the derived processes are documented in Appendix E, Table E.1.

Compared to the original FBS, there is an increase in the capture of the reformulations. The increase is most noticeable for Reformulation II (behaviour) for this protocol.

The remaining 8 % contain processes like B^i ⇝ S^e and B^i ⇝ S^i. Figure 4.7 contains both examples, the derived process from segment 40 to 41 is an example of B^i ⇝ S^e. The processes from links between segments 42 ⇝ 43 and 40 ⇝ 43 are examples of B^i ⇝ S^i. In the first round of coding, segment 40 "the sledge manages to keep level by having quite a wide base" was coded as B^i; in the second round it was coded as S^i. The final arbitrated code was B^i. It should contain two parts—the behaviour part of "keep level" and the structural part "wide base". Segment 43 "unlike the set of skis" was also one of those codes about which there was disagreement (B^i and S^i). The final code was S^i but by carefully examining the context the analogy of "unlike" was both structure and behaviour; the structure analogy was "wide base" against "narrow", the behaviour analogy was "force in the middle" versus on one leg. Figure 4.8 illustrates a proposed refinement of the segments from segment 40 to 43 together with the codes and linkograph. The first column in the table contains the segments with an alphabetic suffix added to those subdivided segments. The links were updated so that they corresponded to the processes of the ontology.

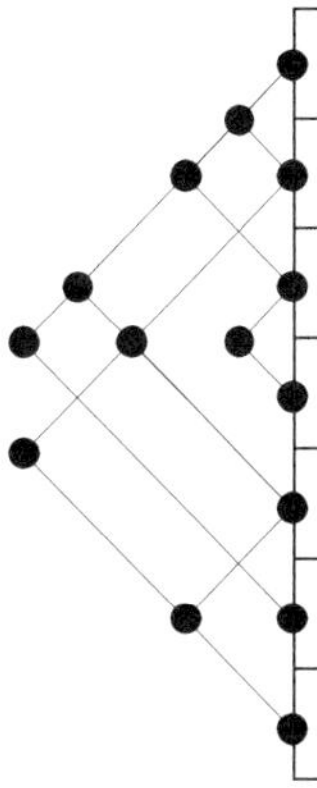

40a	S^i	J: the sledge by having quite a wide base
40b	B^i	J: manages to keep level
41a	Se^i	A:[interpret and expecting the "wide base" structure of a sledge]
41b	S^e	A: (write: wide base)
42	B^i	J: a main force in the middle
43a	S^i	J: unlike the set of skis [in terms of structure]
43b	B^i	J: unlike the set of skis [in terms of behaviour]

Fig. 4.8 Proposed finer grain of re-segmenting and recoding

The missing processes (8 %) were caused by the lack of experience in using this coding method; this includes choosing the correct granularity and making appropriate links. In this case some of the segments require a finer grain than was used. Further analysis and refinement is likely to resolve these "missing" processes as exemplified above.

4.1.7 Findings in Case One

The FBS ontology denotes fundamental processes of designing that are general enough to embrace almost all design situations. Unlike most coding schemes, supported by available protocol analysis software that allows overlapping of codes, the ontological approach requires precise discernment of one code per segment. This clear distinction converts the protocol into unambiguous segments; it quantifies the amount of effort spent in relation to function, behaviour or structure. The links not only provide a structural view of the processes but also locate the dominant codes and the frequency of each design transformation process. The nested representation of links, the linkograph, together with the FBS-coded segments provide an opportunity to look into the design protocol not in a linear manner, but as a network of processes.

The use of the FBS ontology has been able to capture the design semantics of this protocol. Of particular interest is that formulation/reformulation is the largest activity in terms of events and that the vast majority of reformulation is concerned with behaviour and structure. This maps well to our qualitative understanding of this session—generating ideas by analogy. The FBS-based coding scheme accounts for 92 % of all designing activities in this protocol. Later protocol studies brought the percentage of design activities accounted for closer to 100 % (Jiang 2012).

4.2 Case Two: Face-to-Face Versus Computer-Mediated

In this case, data were obtained from the CRC for Construction Innovation project (Kan and Gero 2008b). In that project in vitro studies were conducted with five pairs of designers. One pair was selected for this study. The most creative face-to-face session, judged by the design outcome, was selected for analysis. In the experiments each pair was asked to collaborate in three different settings: face-to-face, Internet Group Board and a 3D virtual world. Group Board is a shared drawing-board environment, in which designers could communicate via the Internet in remote locations. The version used was a commercial version. The 3D virtual world is an extension of Active World and includes video contact, a shared whiteboard and an object viewer/insert of building objects. The design tasks were to generate conceptual designs; each session lasted for 30 min and began with briefs of similar complexity. The same site was used in all the sessions. In the face-to-face session the designers were asked to design a contemporary art gallery. The brief for the Internet Group Board sessions was an architectural library for the university; and the brief for the 3D virtual world was a dance studio. Prior to the computer-mediated experiments, there were training sessions to acquaint the participants with the operations of the different environments. For the analysis, all the sessions were video-recorded. Figure 4.9 shows a frame of the digital recording of the 3D-world session; two of the channels recorded their screens. Their full resolution screens were also captured at three frames per second, illustrated at top left and bottom right in the figure.

The Group Board session was not studied, because Maher et al. (2006) showed in their analysis that in the Group Board session, designers behaved either similarly to the face-to-face session or in the mid-ground between the face-to-face and 3D virtual-world session.

Fig. 4.9 Digital recording of the 3D-world session

4.2.1 *Qualitative Analysis*

A1 and A2 are used to represent the two participants in this section. In all the sessions A1 seemed to take the leadership role and made decisions; he drew most of the sketches in the face-to-face session and organised most of the activities in the 3D virtual-world session.

Face-to-Face Session

The face-to-face session can be divided into four stages or episodes, based on the design activities. In the first episode the two designers dealt with the brief and site (about 3.5 min); in the second episode they analysed, planned and developed concepts in the plan (Fig. 4.10, about 9 min); in the third episode they developed the 3D form in elevation (Fig. 4.11, about 9 min); and in the final episode they worked on the layout in the plan until the end (Fig. 4.12, 8.5 min), but they did not finish it within the 30 min allocated for the session.

They started by analysing the site. Both of them knew the site, although A2 got the orientation wrong. In the second episode, after analysing the site A1 suggested the location of the main approach and started drawing. A2 suggested the location of the service entrance. Issues such as an icon to capture attention, internal and external relationships, and permeability were discussed. Also, the location for the main exhibition, the back of the house, and a merchandising area were suggested (Fig. 4.10). Afterwards A2, using another sheet (Fig. 4.11), suggested the elevations and the idea of a "ribbon", but A1 provided a counter proposal—the "hole in the middle" idea—by drawing an isometric view (Fig. 4.11). In the third episode, they combined the two ideas and further developed it in the plan (bottom left of Fig. 4.11). Eventually A1 drew the 3D form (right of Fig. 4.11). In the last episode, they tried to resolve the dimensions and constraints of the design, which involved calculations (Fig. 4.12).

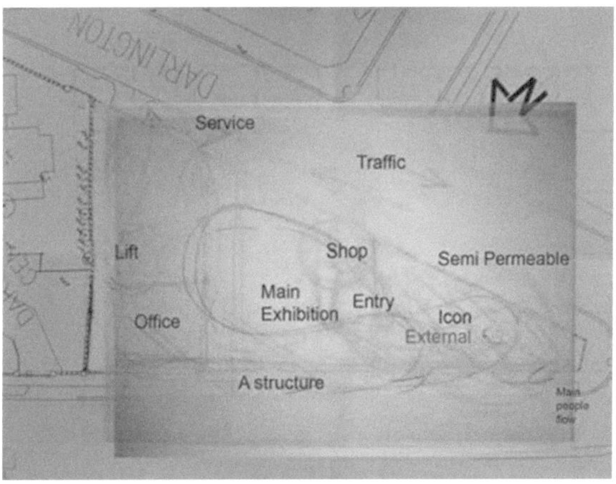

Fig. 4.10 First two sheets of plans in the first and second episode

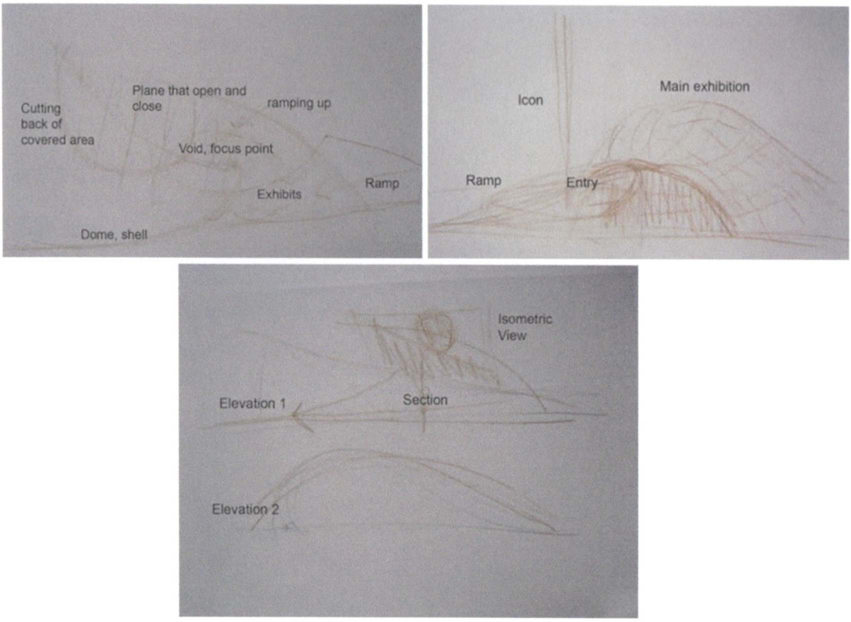

Fig. 4.11 Plans, section and elevation in the third episode

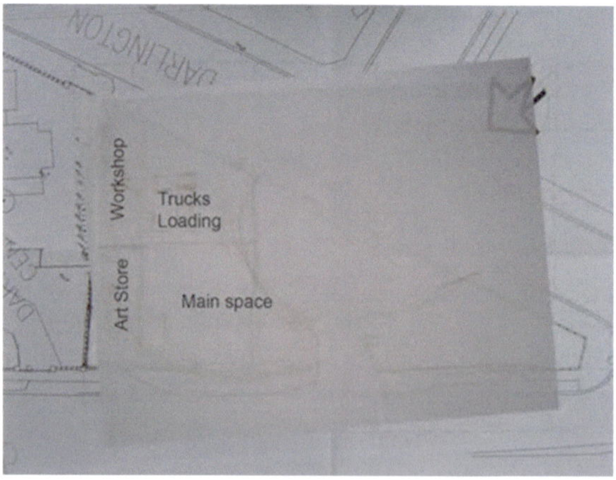

Fig. 4.12 Plan in the last episode

3D-World Session

In the 3D-world session, the stages were not as well defined as in the face-to- face session; they spent less than 2 min with the brief before exploring and making objects. This session can be characterised as "designing through making". Sometimes they subdivided the tasks and worked individually. They were given pre-defined elements—space, slab, wall, column and beam—in various sizes. They decided to start with the biggest space element to represent the "largest" spaces, the

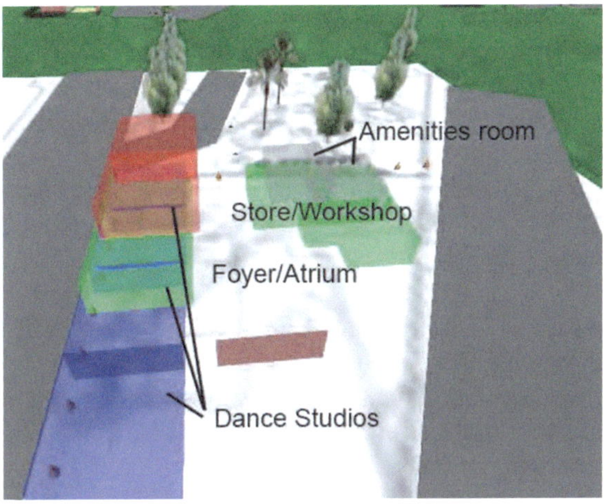

Fig. 4.13 Final design of the 3D-world session

four studios. At around 12 min, they discovered they could not have all the studios on one level because of the site coverage constraint. A1 decided to stack them and create an atrium to join them together. They tried to further develop this concept to accommodate the requirements (Fig. 4.13). A1 repeatedly went to pick up those space objects of relevant size for A2 to arrange on the site.

They did not finish the design and left out elements such as connecting bridges and some functional spaces. Besides designing, time was spent on design support activities, such as discussing what elements were available and organising what to do. Also, time was spent on the technical aspects of learning how to do things, such as changing the colour of the blocks, how to "fly", and how get out when "trapped inside" those blocks.

Comparing the Two Sessions

This section qualitatively compares the two sessions. In terms of the design outcome, as seen in Figs. 4.11 and 4.13, the styles of the designs were very different. The design of the face-to-face session was free-form and organic, while the design of the 3D-world session was orthogonal.

In terms of the process, in the face-to-face session the design process was closely coupled, while in the 3D virtual-world session the process was loosely coupled. In the 3D-world session the designers tended to work more individually. Working individually led to the issue of the sense of presence.

The designers made used of the avatar in the 3D virtual-world to detect the other party; also they used the web cam. Issues about the sense of presence often appeared in the protocol in the 3D-world session, such as: "…you're not looking my way anyway", "the camera is not directed at you", and "I can't see you though…, I don't know where you are…". Also, they lost the ability to gesture. The design actions occurred through interaction with keyboard and mouse. In the face-to-face session they relied on gesturing to communicate; they gestured paths, shapes and circulations. They also used gestures to signal turn-taking with the drawing.

The length of the verbal protocol in the face-to-face session was not only longer but also more concentrated on designing. There was more non-design activity in the 3D-world protocol. For example, communication regarding the software: "how do change the colour", communication related to the location of each other: "where are you?", communication regarding the ownership of objects "now you've taken it away", and other social communication: "it's superman". The amount of time during which they remained silent was also longer in the 3D-world session. Even in design-related communication, they were more concerned with achieving tasks; this can be explained by the limitation of the software, for example: "I pick it up…", "we bring it across…", and "… because it is filled by blocks".

In terms of idea development, the designers developed more design ideas in the face-to-face setting than in the 3D virtual-world setting. There were a number of concepts besides organising the space, such as: "make it a journey of discovery", "the ribbon idea", "hole in the middle", "ramping … this whole platform" and "dropping into the centre". These ideas co-evolved in the problem and solution space and it is difficult to pinpoint what triggered these ideas. What was observed matched what Finke et al. (1992) described as many processes collectively setting the stage for creative insight and discovery. In the 3D-world session, the main idea was the "atrium". Although there were fewer ideas, the designers still switched between the problem and solution space.

4.2.2 Ontological Coding of Both Sessions

The design protocols were segmented according to the situated FBS ontology, as described in Sect. 3.4.3 of Chap. 3.

Table 4.8 is an extract of the coded protocol of the face-to-face session. The segmentation is based on the FBS ontology. Segment 13 was a response triggered by attending to the requirement of "size" in the brief. The architect was expecting to obtain structural data by pulling data from the brief, therefore it was coded as "S^i". Segments 14 and 16 were read from the brief so they were coded as "R". Segments 15 and 17 represented the interpretation of the behavioural and structural requirements, so they were coded as "B^i" and "S^i" respectively. Segment 18 recalled a famous museum by Frank Gehry. From the context it seemed to refer to the structure, expecting the gallery to look like "the Guggenheim". This gave a new meaning to the understanding of the design. In segment 15 the architect interpreted the gallery as "typical" when referring to the "permanent and temporary" collection

Table 4.8 An extract of the coded protocol of the face-to-face session

Segments	Protocol	Code
13	Okay hang on, it's talking about sizes here	S^i
14	So … (read brief) permanent and temporary	R
15	Typical	B^i
16	(read brief) Permanent collection is 200 and 50 m hanging space	R
17	50 m hanging space!	S^i
18	This is the Guggenheim	Se^i

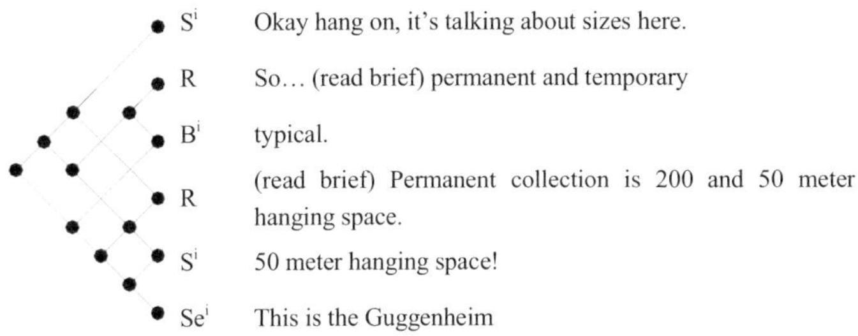

Fig. 4.14 Example of linkograph in relation to protocol

space; this interpretation was changed in the light of the current situation. Figure 4.14 shows the coded protocol together with the linkograph constructed by discerning how segments are related.

The coding of the first 11 min of both sessions are used here. There were 205 segments and 95 % of them contained FBS codes in the face-to-face session. There were 125 segments in the 3D-world sessions and 51 % of them contained FBS codes. The low percentage in the 3D-world was a result of learning how to do things, especially in the beginning, like "how to fly", "how to get out", and "how to change colour". Actions of the mouse and keyboard in the 3D-world session were not segmented, so no external world actions were coded. Regardless, there was a high percentage of structure, which corresponded to the "design by making".

There are 595 links in the face-to-face session and 92 links in the 3D-world session. Figures 4.15 and 4.16 compare the percentages of each code of the two sessions in the segments and in the links respectively. In the 3D session many actions and activities were not captured because no verbalisation occurred while the designers were using the keyboard. These actions include the manipulation of objects in the 3D-world.

The segments that are not design related can be coded as "Other" with an "O". These segments can be deleted from the when carrying out only design-related analyses.

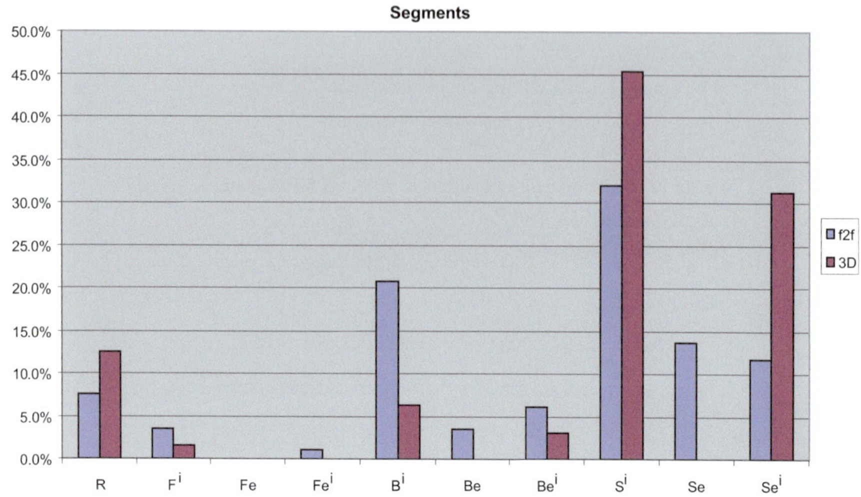

Fig. 4.15 The distributions of codes in the segments of the face-to-face and the 3D-world sessions

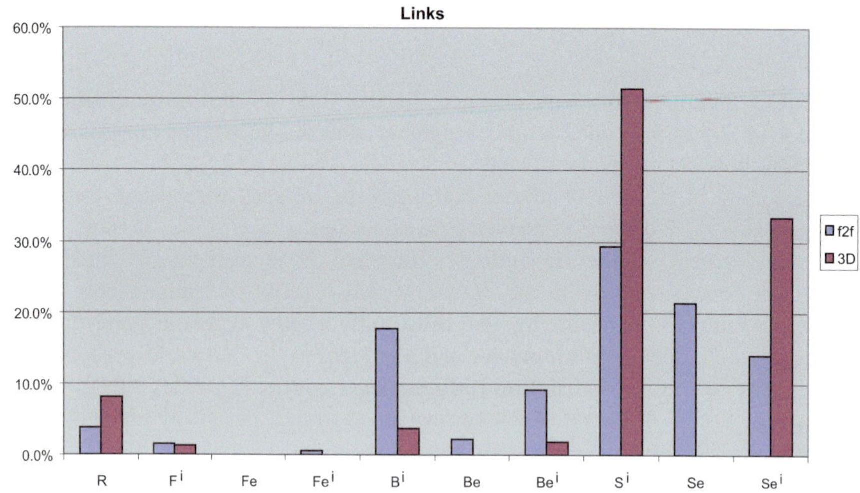

Fig. 4.16 The distributions of codes in the links of the face-to-face and the 3D-world sessions

4.2.3 The Eight FBS Processes of Both Sessions

Processes were derived from the links. If segments coded as O are included some of these processes are meaningless, because some segments do not have an FBS code. Using a similar method to that described in Case One, the derived processes were

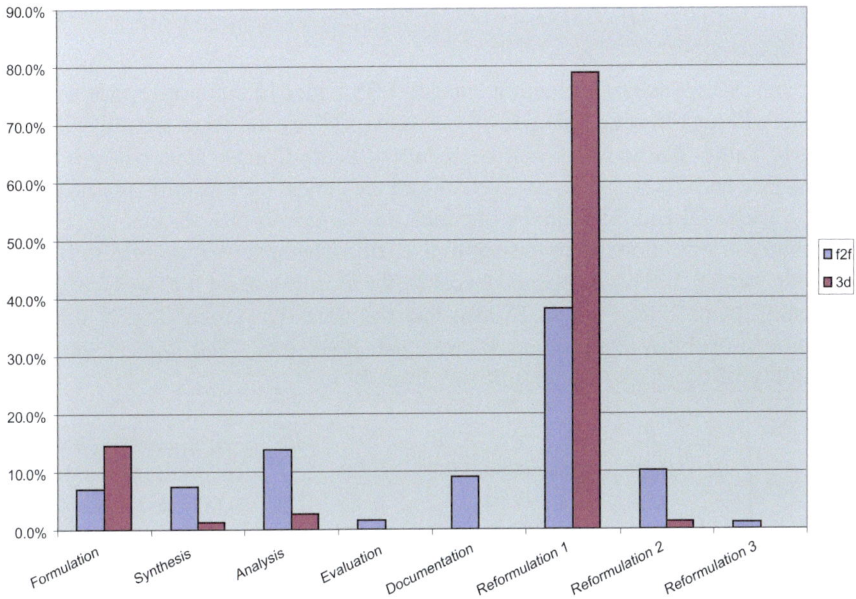

Fig. 4.17 The eight FBS processes of the face-to-face and the 3D-world sessions

grouped into the eight FBS categories. Figure 4.17 compares the percentages of the grouped processes. This shows and compares the distribution of processes but not the quantity of those processes. For example, the 79 % of the type one reformulation process in the 3D-world session has 60 processes but the 38 % in the face-to-face session has 225 processes.

In the face-to-face session, these eight processes add up to about 89 % of all the derived processes, while in the 3D-world they aggregate to about 99 %. Again, the missing processes, especially in the face-to-face session, were caused by the granularity of the segments. There is no documentation process in the 3D-world session, because no external world actions were coded. Also, there is no evaluation process; this is the result of coding only the first 11 min. In this short period, the designers hardly had anything to evaluate but had a considerable amount to formulate. The percentages of all the derived processes are documented in Appendix E, Tables E.2 and E.3.

All three types of reformulations were present in the face-to-face session, but no Reformulation 3 was found in the 3D-world session. Both sessions have a relatively high type one reformulation. The face-to-face session has higher analysis, synthesis and evaluation processes. These matched the qualitative understanding of the sessions. In the 3D-world session the predominant process was the reformulation of structure, the re-making of forms.

4.2.4 Markov Analysis

There are 197 segments containing situated FBS codes in the face-to-face session, so there will be 196 state changes. In the 3D-world session there are only 63 state changes. Tables 4.9 and 4.10 list the occurrences of all those state changes in the face-to-face and 3D-world respectively. From the tables, the transition matrices, P_{f2f} and P_{3d} (Eqs. 4.4 and 4.5), can be obtained. In the face-to-face session, the highest probability of transitions were from S^i to S^i (0.52). The next two were Fe^i to Be^i and S^i, both were 0.5. This happened because the occurrence of Fe^i was rare, only happening twice. From Be^i to S^e also had the transition probability of 0.5. The transition probability from B^e to B^i was also high, 0.43. The highest transition probability of the 3D-world session was from B^i to R.

$$P_{f2f} = \begin{pmatrix} & R & F^i & Fe^i & B^i & Be^i & B^e & S^i & Se^i & S^e \\ R & 0.20 & 0.20 & 0.00 & 0.13 & 0.00 & 0.07 & 0.40 & 0.00 & 0.00 \\ F^i & 0.29 & 0.14 & 0.14 & 0.29 & 0.14 & 0.00 & 0.00 & 0.00 & 0.00 \\ Fe^i & 0.00 & 0.00 & 0.00 & 0.00 & 0.50 & 0.00 & 0.50 & 0.00 & 0.00 \\ B^i & 0.10 & 0.02 & 0.02 & 0.34 & 0.17 & 0.05 & 0.22 & 0.02 & 0.05 \\ Be^i & 0.08 & 0.00 & 0.00 & 0.00 & 0.00 & 0.17 & 0.00 & 0.25 & 0.50 \\ B^e & 0.14 & 0.00 & 0.00 & 0.43 & 0.00 & 0.00 & 0.29 & 0.14 & 0.00 \\ S^i & 0.05 & 0.02 & 0.00 & 0.13 & 0.00 & 0.02 & 0.52 & 0.13 & 0.14 \\ Se^i & 0.00 & 0.00 & 0.00 & 0.14 & 0.09 & 0.05 & 0.14 & 0.23 & 0.36 \\ S^e & 0.04 & 0.00 & 0.00 & 0.33 & 0.04 & 0.00 & 0.33 & 0.19 & 0.07 \end{pmatrix} \quad (4.4)$$

Table 4.9 Occurrence of the sequence of situated FBS codes in the face-to-face session

State	Next State								
	R	F^i	Fe^i	B^i	Be^i	B^e	S^i	Se^i	S^e
R	3	3	0	2	0	1	6	0	0
F^i	2	1	1	2	1	0	0	0	0
Fe^i	0	0	0	0	1	0	1	0	0
B^i	4	1	1	14	7	2	9	1	2
Be^i	1	0	0	0	0	2	0	3	6
B^e	1	0	0	3	0	0	2	1	0
S^i	3	1	0	8	0	1	33	8	9
Se^i	0	0	0	3	2	1	3	5	8
S^e	1	0	0	9	1	0	9	5	2

Table 4.10 Occurrence of the sequence of situated FBS codes in the 3D-world session

State	Next state					
	R	F^i	B^i	Be^i	B^e	S^i
R	1	0	3	0	4	0
F^i	1	0	0	0	0	0
B^i	3	0	0	0	1	0
Be^i	0	0	0	1	0	1
B^e	1	1	0	0	18	8
S^i	1	0	1	1	6	11

$$P_{3d} = \begin{pmatrix} & R & F^i & B^i & Be^i & S^i & Se^i \\ R & 0.13 & 0.00 & 0.38 & 0.00 & 0.50 & 0.00 \\ F^i & 1.00 & 0.00 & 0.00 & 0.00 & 0.00 & 0.00 \\ B^i & 0.75 & 0.00 & 0.00 & 0.00 & 0.25 & 0.00 \\ Be^i & 0.00 & 0.00 & 0.00 & 0.50 & 0.00 & 0.50 \\ S^i & 0.04 & 0.04 & 0.00 & 0.00 & 0.64 & 0.29 \\ Se^i & 0.05 & 0.00 & 0.05 & 0.05 & 0.30 & 0.55 \end{pmatrix} \tag{4.5}$$

Probability Vector

Putting P_{f2f} (Eq. 4.4) into Eq. 3.2 and solving the equations, the probability vector for the face-to-face session is:

$$a_{f2f} = \begin{pmatrix} R & F^i & Fe^i & B^i & Be^i & B^e & S^i & Se^i & S^e \\ 0.07 & 0.03 & 0.01 & 0.21 & 0.06 & 0.04 & 0.32 & 0.12 & 0.14 \end{pmatrix} \tag{4.6}$$

This distribution is similar to the distribution of codes in segments. With a large number of FBS segments, 32 % of the codes are predicted to be S^i, 21 % of the codes are predicted to be B^i, 14 % of the codes are predicted to be S^e, 12 % of the codes are predicted to be Se^i, 10 % of the codes are predicted to be the combination of Be^i and B^e, 7 % of the codes are predicted to be R and 4 % of the codes are predicted to be the combination of F^i and Fe^i.

Similarly, putting P_{3d} (Eq. 4.5) into Eq. 3.2 and solving the equations, the probability vector for the 3D-world session is:

$$a_{3d} = \begin{pmatrix} R & F^i & B^i & Be^i & S^i & Se^i \\ 0.10 & 0.02 & 0.06 & 0.03 & 0.46 & 0.33 \end{pmatrix} \tag{4.7}$$

This distribution is also not far from the distribution of codes in segments. The Markov probability predicts fewer R and B^i codes. The analysis predicts S^i and S^e will occupy nearly 80 % of the codes with a large number of segments. Comparing this to α_{f2f} (Eq. 4.6), the 3D-world session is predicted to have more structure codes.

First Passage Times

Using P_{f2f} and α_{f2f} together with Eqs. 3.4 and 3.5 the mean first passage times for the face-to-face session can be obtained:

$$M_{f2f} = \begin{pmatrix} & R & F^i & Fe^i & B^i & Be^i & B^e & S^i & Se^i & S^e \\ R & 13.5 & 32.5 & 102.1 & 5.9 & 16.4 & 26.6 & 3.7 & 11.7 & 8.7 \\ F^i & 12.2 & 34.2 & 87.2 & 5.3 & 12.6 & 27.3 & 5.4 & 11.7 & 8.3 \\ Fe^i & 18.0 & 43.4 & 107.5 & 7.0 & 9.4 & 27.1 & 3.6 & 10.0 & 6.5 \\ B^i & 15.8 & 40.8 & 102.6 & 4.8 & 13.3 & 26.2 & 4.6 & 10.8 & 7.5 \\ Be^i & 16.7 & 42.7 & 106.7 & 6.0 & 16.5 & 24.0 & 5.2 & 8.3 & 4.3 \\ B^e & 15.3 & 41.3 & 105.2 & 4.3 & 16.0 & 27.9 & 4.2 & 10.1 & 8.1 \\ S^i & 17.2 & 42.0 & 106.3 & 5.9 & 16.9 & 28.3 & 3.1 & 9.7 & 6.8 \\ Se^i & 18.1 & 43.4 & 106.8 & 5.6 & 14.9 & 26.9 & 4.9 & 8.4 & 4.8 \\ S^e & 17.3 & 42.6 & 106.0 & 4.8 & 15.6 & 28.1 & 4.0 & 9.3 & 7.1 \end{pmatrix} \tag{4.8}$$

In general it takes longer to move from F to S^e than from B to S^e. It is unexpected that the first passage time from Be^i to S^e is shorter than from Se^i to S^e. Further investigation is required to explain this.

Similarly, using P_{3d} and α_{3d} together with Eqs. 3.4 and 3.5 the mean first passage times for the 3D-world session are:

$$M_{3d} = \begin{pmatrix} & R & F^i & B^i & Be^i & S^i & Se^i \\ R & 9.6 & 59.6 & 14.8 & 65.1 & 2.3 & 6.5 \\ F^i & 1.0 & 60.6 & 15.8 & 66.1 & 3.3 & 7.5 \\ B^i & 4.5 & 60.0 & 18.0 & 65.5 & 2.7 & 6.9 \\ Be^i & 15.7 & 62.7 & 24.5 & 30.3 & 5.4 & 2.0 \\ S^i & 13.9 & 57.3 & 23.8 & 62.8 & 2.2 & 4.2 \\ Se^i & 13.7 & 60.7 & 22.5 & 58.6 & 3.4 & 3.0 \end{pmatrix} \tag{4.9}$$

4.2.5 Findings and Discussion for Case Two

Maher et al. (2006) concluded that the characteristics of the design process are quite different in sketching and 3D virtual environments. The preliminary results confirm that by showing the difference in the distribution of processes. Also, the amount of reformulation processes were much higher in the face-to-face session, which corresponded to the designers having developed more design ideas in the face-to-face setting than in the 3D virtual-world settings.

In order to perform a more rigorous comparison between the face-to-face and the 3D-world session, the protocol of the 3D-world session will need re-segmenting to include actions such as copying objects, changing object's colour, and navigating/moving in the 3D world. It will be helpful to have the logs of the designers' interactions for consultation. Also, the length of the protocol studied in the 3D world session was too short to cover a greater variety of different types of processes.

The study of the interactions among the FBS classes and processes in a set of design sessions rather than a single session can help to deepen the understanding of designing in a 3D-world situations, which can then inform the development of tools that aid designing in this environment.

4.3 Case Three: Statistical Exploration of the Effects of Education on Design Cognition

In this case we present an exploration by Williams et al. (2011) in which they employed hypothesis testing using FBS coding to study the effects of education on design cognition. Along with Student's t test (as explained in Chap. 3) other statistical techniques such as the Shapiro–Wilk W test was used. The principle of setting up the null hypothesis and getting the p value was the same. In this case, the authors report on progress of a longitudinal study on the impact of design education on students' design thinking and practice. Data gathered from two experimental sessions (conducted before and after the students' introductory design course) are analysed to identify changes in their design cognition regarding the FBS-coded design issues and the syntactic design processes.

4.3.1 Participants and the Experimental Setup

The participants of this case were second-year Mechanical Engineering students enrolled in a design course—Engineering Design and Economics—that aimed to expose them to engineering design and design methodologies at an early stage in their professional development. The 3-credit design course is centred on active-learning opportunities that allow students to apply their learning in engineering design. Classroom meetings are typically devoted to hands-on team based activities, which range from product dissections (internal combustion engines, air compressors, electric drills, disposable cameras, etc.) to various speculative design scenarios. These activities provide an opportunity for the instructor to perform individual mentoring and instruction. In addition to these in-class activities, student design teams work together out-of-class on a semester project where they design a novel consumer product.

The data collected of this study represents the beginning of the students' formal design education and experience. At this stage, the students' design education is

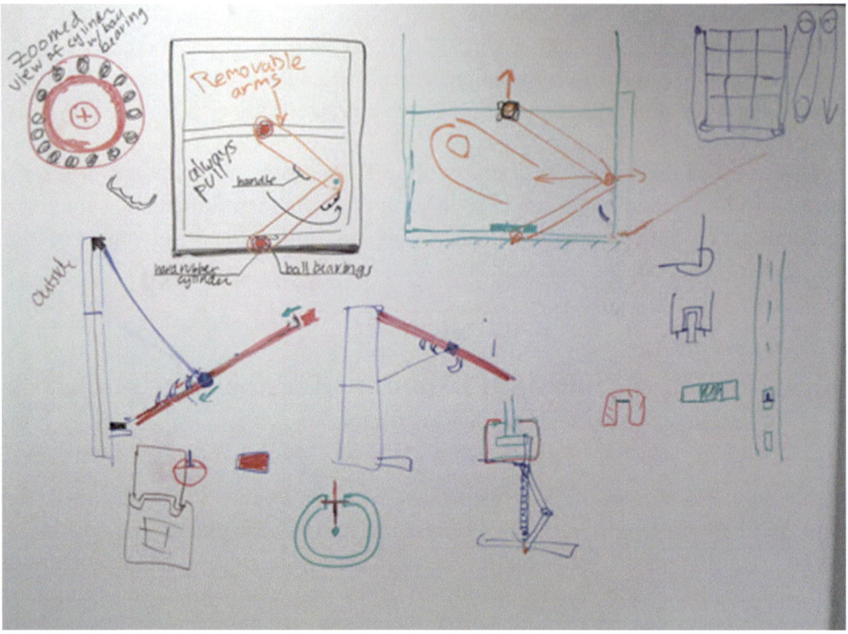

Fig. 4.18 Participant white-board sketch examples

limited to a 4-week "Introduction to Engineering Design" module in their first-year introductory engineering course. In order to make sure students are novices, those with significant design experience (either professionally or through prior academic experience), were screened through a preliminary interview and were not selected as participants.

Participants were asked to attend two out-of-class experiments: one at the beginning of the fall semester of their sophomore year; the other in the middle of the spring semester of the same academic year. In these experiments, pairs of students worked together at a white board to solve a speculative design task. A total of 28 students (16 in fall 2009 and 12 in spring 2010) participated the design session. Figure 4.18 shows examples of typical sketches produced during the design session.

Participants were paired up and given 45 min to come up with a design solution that meets the requirements in the provided brief. The entire design sessions were audio and video recorded for analysis. Two digital camcorders were used, one recording the whiteboard and the other recording the participants. Each participant had individual remote microphones to ensure the recording quality of their conversation. These raw data were segmented according to the FBS design issues.

4.3.2 Results and Findings of Case Three

Design Issues

The distributions of design issues before and after the introductory design course are illustrated in Fig. 4.19, with descriptive statistics reported in Table 4.11.

In both data sets it is observed that students spent the majority of their cognitive effort discussing design structure (37–40 %), followed by behaviour from structure (30–32 %). These two design issues accounted for two-thirds of their cognitive effort. Much less cognitive effort was spent on the design issues of description (9–15 %), expected behaviour (6–11 %), function (2–7 %), and requirement (2–3 %). The variations between before and after being exposed to the design course have been identified for each design issue. The percentages of their cognitive effort related to design function and design description have increased approximately 5 and 6 % respectively, whereas the percentages for all the other design issues have decreased. JMP 9.0, a statistical software package, was utilized to identify any statistical differences in the percentages of cognitive effort related to individual design issues before and after the course. Statistically significant differences were

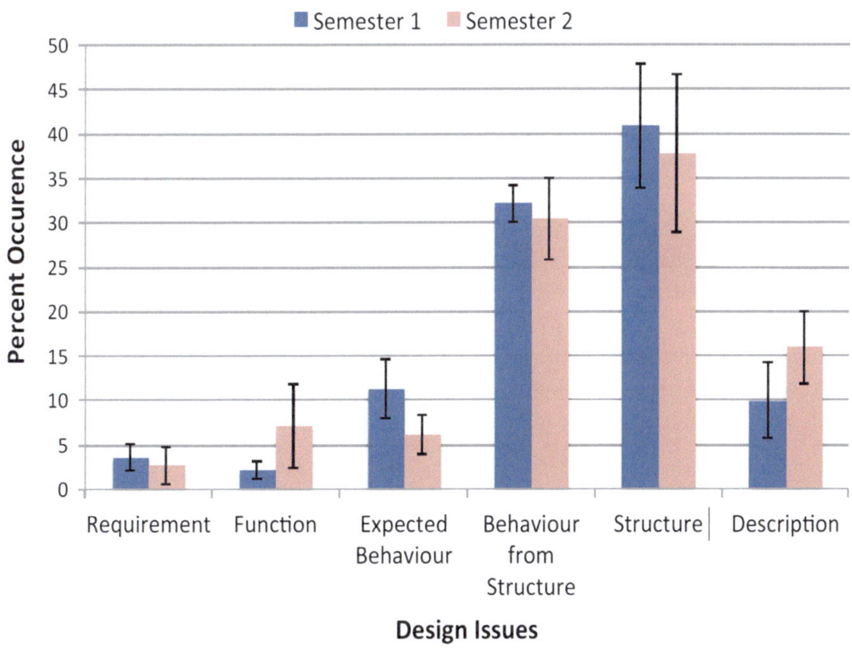

Fig. 4.19 Percent occurrences of design issues before (Semester 1) and after (Semester 2) exposure to design teaching

Table 4.11 Means and standard deviations of design issues before (Semester 1) and after (Semester 2) exposure to design teaching

Issue distribution	Semester 1 (n = 8) mean (SD)	Semester 2 (n = 6) mean (SD)
Requirement	3.62 (1.46)	2.70 (2.10)
Function	2.16 (1.07)	7.10 (4.67)
Expected behavior	11.31 (3.32)	6.13 (2.22)
Behavior from structure	32.13 (2.12)	30.37 (4.58)
Structure	40.88 (7.01)	37.75 (8.83)
Description	9.91 (4.22)	15.93 (4.10)

assumed at a significance level (α) of 0.05. The normality assumption was tested for each design issue using the Shapiro–Wilk W test. Only the percentage of cognitive effort on the design issue of function rejects the null hypothesis, which states that the data are from the normal distribution. Therefore, the Wilcoxon rank sum test, a non-parametric statistical analysis, was used for the design issue of function, whereas two-sample t tests were used for the rest. The results for these statistical analyses are reported in Table 4.12.

The statistical analyses indicate that there are significant differences regarding the percentages of cognitive effort in the three design issues of function, expected behaviour, and description between the two semesters. In semester 2, students were more engaged in discussions related to the design function, which is the teleology of their design solution. Specifically, many students attempted to jointly optimize their design for two criteria: design for the users (patients) and design for the customer (rehabilitation institute). Students intended to solve the initial problems with their design, yet also aimed to add or modify the functionalities of their design based on criteria such as safety, security, or possibilities to failure.

The percentage of cognitive effort on the design issue of expected behaviour significantly decreased from Semester 1 to Semester 2, after taking a design course. It is possible that this cognitive change could be caused by the design course's 3-week focus on problem formulation and functional decomposition. In these portions of the course students are taught to scope a design problem by identifying customer needs, transforming them into target specifications, completing a needs-metrics matrix and formulating function structures.

Table 4.12 Design issues comparisons between Semester 1 and Semester 2 statistical analysis results

Design issue	t(z) statistics	p value
Requirement	−0.925	0.137
Function	2.904	0.003**
Expected behaviour	−3.495	0.004**
Behaviour from structure	−0.879	0.409
Structure	−0.717	0.490
Description	2.685	0.021*

*$p < 0.05$; **$p < 0.01$

The percentage of cognitive effort on the design issue of description significantly increased after taking a design course. This increase could be due to the introductory design course as well—a major learning goal of the course is effective oral and written communication of design outcomes. As a result, students might be more confident in explaining their designs.

No significant differences in the percentages of cognitive effort on the design issues of requirement, behaviour from structure, and structure between the two semesters were identified.

Syntactic Design Processes

The syntactic design process distribution was computed to identify differences between before and after an introductory design course. The occurrences of syntactic processes for the two semesters are illustrated in Fig. 4.20, with descriptive statistics reported in Table 4.13.

The majority of students' cognitive effort was expended on reformulation I (29–33 %) and analysis (28 %), which accounted for almost two-thirds of their cognitive effort. Much less cognitive effort was spent on the design processes of documentation (10–16 %), evaluation (7–10 %), synthesis (6–8 %), reformulation II (4–6 %), reformulation III (1–4 %), and formulation (1 %). Some differences

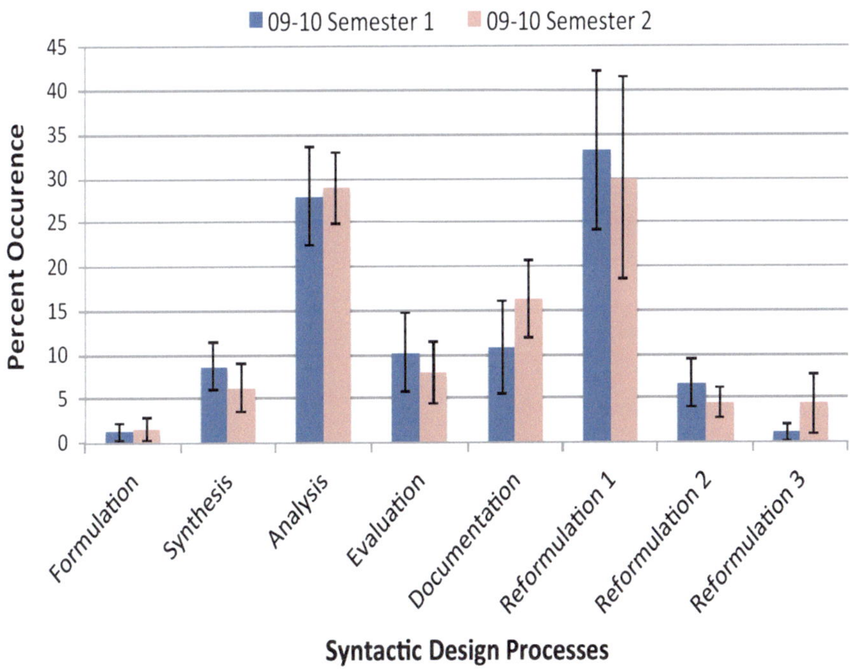

Fig. 4.20 Percent occurrences of syntactic design processes before (Semester 1) and after (Semester 2) exposure to design teaching

Table 4.13 Means and standard deviations of design issues before (Semester 1) and after (Semester 2) exposure to design teaching

Syntactic process distribution	Semester 1 (n = 8) mean (SD)	Semester 2 (n = 6) mean (SD)
Formulation	1.32 (1.01)	1.65 (1.30)
Synthesis	8.70 (2.66)	6.28 (2.70)
Analysis	28.00 (5.55)	28.95 (4.08)
Evaluation	10.23 (4.44)	7.96 (3.57)
Documentation	10.77 (5.31)	16.30 (4.43)
Reformulation I	33.18 (9.00)	29.98 (11.42)
Reformulation II	6.72 (2.70)	4.45 (1.69)
Reformulation III	1.06 (0.92)	4.38 (3.43)

were identified between the two semesters for each design process. A large increase in the percentages of students' cognitive effort for documentation (+6 %) and reformulation III (+3 %) were identified, whereas a slight increase was identified for formulation (0.33 %) and analysis (0.95 %). The cognitive effort for design processes of synthesis, evaluation, and reformulation II have decreased approximately 2 %, whereas, reformulation I decreased 3 %.

Two-sample t-tests were used to identify significant statistical differences in the percentages of cognitive effort related to individual design processes before and after the course, as shown in Table 4.14. As the percentage of cognitive effort on the design process of reformulation III did not follow the normal distribution, the Wilcoxon rank sum test was used.

The results indicate that there is a significant difference of the percentage of cognitive effort on the design process of reformulation III. The design process of reformulation III represents changes in the design function when the actual behavior is evaluated to be unsatisfactory (Gero and Kannengiesser 2004). The result reveals that the students were more engaged in the design process of modifying design

Table 4.14 Syntactic design processes comparisons between Semester 1 and Semester 2 statistical analysis results

Syntactic process	t(z) statistics	p value
Formulation	0.506	0.624
Synthesis	−1.666	0.124
Analysis	0.368	0.718
Evaluation	−1.059	0.155
Documentation	−0.567	0.584
Reformulation I	−0.567	0.584
Reformulation II	−1.931	0.077
Reformulation III	2.070	0.038*

$^{*}p < 0.05$

issues of function based on their analysis of design structures. As explained in Sect. 4.1, the significant increase of discussions related to the design function in semester 2 provides some evidence of such result.

A smaller pattern of difference between the two semesters was identified for the design process of reformulation II. The design process of reformulation II addresses changes in the design behaviour when the actual behaviour of a design structure is evaluated to be unsatisfactory (Gero and Kannengiesser 2004). This indicates that the students displayed a tendency of decreased engagement in the design process of modifying design issues of expected behaviour based on their analysis of design structures. The result is supported by the significant decrease of discussions of expected behaviour reported in Sect. 4.3.1.

The results highlight that there were significant differences in students' design cognition between before and after an introductory design course. Particularly, the design issues of function and description have significantly increased, while expected behaviour significantly decreased. The syntactic design process of reformulation III significantly decreased as well.

4.4 Case Four: Metacognition of Designing—Problem-Solution Index

Metacognition usually refers to cognition about one's own cognition (Flavell 1979). Dunlosky and Metcalfe (2009) give examples of using metacognition to improve our daily lives such as writing a note when it is essential to remember something. A large part of their model of metacognition concerns monitoring and controlling cognitive memory and processes. In the area of design cognition, it is often modelled as a search process across two notional design "spaces" of problem and solution. Problem (or Project) Based Learning has been used in many architectural school as a pedagogical approach for student to learn design, like the Department of Architecture, University of Newcastle, New South Wales, Australia, Technical University of Delft (TUDelft), Netherlands and the Department of Architecture and Civil Engineering, City University of Hong Kong. Problem-based learning is increasingly used in engineering and more recently in computer science education. We consider metacognition of designing, in a narrow sense, as the meta-level structure over the cognitive processes behind design problem and solution spaces; in other words how problem or solution focus is organized in the design cognitive process. Similar to case 4, we would like to explore if education in a particular domain will have any influence in the students' metacognition. In this section we use a study by Jiang et al. (2012) to illustrate the potential of using the FBS ontology for this kind of design research inquiry.

4.4.1 FBS Design Issues Mapping and Problem-Solution Index

In the FBS ontology, problem formulation mainly involves reasoning about requirement, function and expected behaviour, while reasoning about structure and behaviour from structure are related to artefacts as a solution to the formulated problem The ontologically-based design issues can then be categorized into problem-focused and solution-focused design issues as shown in Table 4.15.

The problem–solution (P–S) index is proposed as a ratio measurement, computing the ratio of the sum of the design issues concerned with the problem space to the sum of those related to the solution space, Eq. 4.10.

$$\text{P–S index} = \frac{\sum (\textit{Problem-related issues})}{\sum (\textit{Solution-related issues})} = \frac{\sum (R, F, Be)}{\sum (Bs, S)} \tag{4.10}$$

The P–S index value quantifies the relative focusing on problem or solution. When the P–S index equals one, it indicates that equal cognitive effort has been spent in the problem and solution spaces. We define a design session with a P–S index larger than 1 as one with a problem-focused metacognitive designing style, and a session with a P–S index value less than or equal to 1 as one with a solution-focused metacognitive design style.

4.4.2 The Experiment

24 final-year undergraduate design students participated in this study, 12 from Industrial Design (ID) and 12 from Mechanical Engineering (ME). Two participants, either from the same discipline or different ones, were paired to work collaboratively in two conceptual design tasks. The combination of design teams according to disciplines were: ID, ME and Mixed teams. The design tasks were used based on Keinonen's (2006) taxonomy of product development concepts and visionary concepts. The first task was to design a coffee maker (CM) for the existing market. It simulated a typical initial stage of a normal new product development (NPD) process. Designers were expected to consider practical factors

Table 4.15 Mapping FBS design issues onto problem and solution spaces

Problem/solution space	Design issue
Problem space = Problem-focused design issues	Requirement (R) Function (F) Expected behaviour (Be)
Solution space = Solution-focused design issues	Behaviour from structure (Bs) Structure (S)

related to a NPD project, e.g., market and user analysis, supporting technology and resources. The second task was to design a next-generation personal entertainment system (PES) for the year 2025. Task PES had a very limited amount of determined/unalterable factors. Designers were expected to use design concepts as a tangible means to explore future scenarios.

4.4.3 *Results and Findings*

The results present below suggest that industrial design student teams have a metacognitive design style that is more focused on the design problem than mechanical engineering student teams. This can be seen in the descriptive statistics where they had high frequency of function and expected behaviour issues; a higher value of P–S index; and a significantly higher P–S index value in the first half of the design sessions.

Descriptive Statistics

Each design session's occurrences of design issues were normalized by dividing them with the total number of design issues in that session. The normalized design issue distributions are shown in Table 4.16 and Fig. 4.21.

Table 4.16 Normalized distribution of design issues (%)

Groups	Design issues					
	Requirement (R)	Function (F)	Expected behavior (Be)	Behavior from structure (Bs)	Structure (S)	Description (D)
ID CM						
Mean	0.9	23.6	15.6	20.7	20.3	19.0
SD	0.4	2.8	4.2	3.8	1.7	6.7
ID PES						
Mean	1.5	28.0	23.1	15.7	13.2	18.7
SD	0.6	3.9	2.9	2.4	3.3	4.4
Mix CM						
Mean	0.9	17.9	12.7	27.4	21.5	19.6
SD	0.4	7.9	1.0	6.6	3.0	2.3
Mix PES						
Mean	1.5	17.3	14.4	27.2	18.0	21.6
SD	0.5	2.9	3.6	8.4	4.5	5.2
ME CM						
Mean	1.8	11.4	13.5	28.0	28.3	16.9
SD	0.4	6.0	3.7	3.2	8.2	5.4
ME PES						
Mean	1.1	12.1	15.6	31.2	19.8	20.1
SD	0.4	2.9	6.2	7.2	2.9	7.0

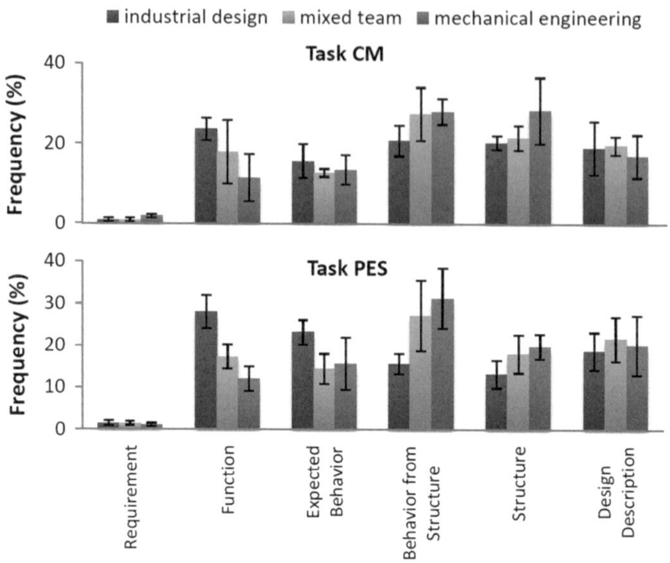

Fig. 4.21 Distribution of design issues (%)

ID sessions had the highest percentages of function issues followed by the Mixed sessions and then the ME sessions. ID sessions also had higher percentages of expected behaviour issues than these other two groups. ME sessions had the highest percentages of the solution-related issues and ID sessions had the lowest ones. For inter-task comparisons, Task PES tended to have more function and expected behaviour issues than Task CM, whereas the percentage of structure issues in Task CM was higher than that in Task PES.

Problem-Solution Indexes of the Whole Sessions

The values of P–S index for each design session are shown in Table 4.17. The problem-focused sessions are highlighted by bold fonts. The results are also aggregated and charted in Fig. 4.22, against a line at the value of 1.00 for the P–S

Table 4.17 Values of P–S index

Groups	Value of P–S index for each team				Mean	SD
	1	2	3	4		
ID CM	0.90	**1.01**	**1.13**	0.88	0.98	0.11
ID PES	**2.04**	**2.32**	**1.74**	**1.40**	1.88	0.39
Mix CM	0.95	0.36	0.89	0.53	0.68	0.28
Mix PES	0.48	0.76	0.77	**1.01**	0.76	0.22
ME CM	0.28	0.34	0.55	0.80	0.49	0.23
ME PES	0.56	0.75	0.46	0.54	0.58	0.12

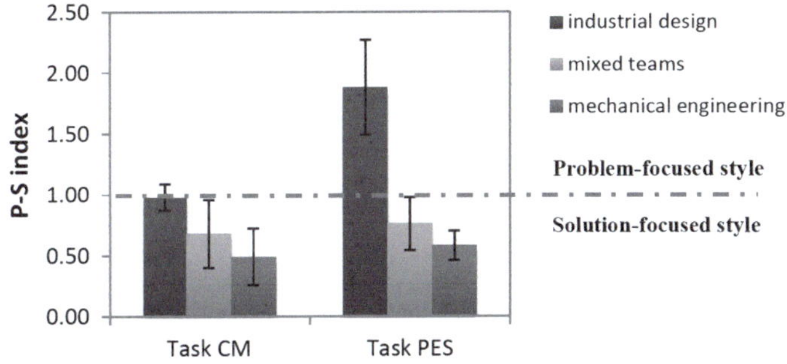

Fig. 4.22 Aggregated P–S index values and metacognitive design style

index signifying the boundary between problem-focused and solution-focused metacognition design styles.

ID PES sessions had significant higher P–S index values than other sessions, demonstrating a strong tendency of focusing on problem-related issues. The P–S index value of the ID CM sessions are around the threshold of problem-solution division. ID teams generally focused more on the design problem than did the mixed and ME teams.

Problem-Solution Indexes in Two Halves

Dividing the design sessions into two halves provides a more nuanced basis to study metacognition—how participants' cognitive effort occurs over time within a design session. Figure 4.23 compares the P–S index values of the two halves with the respective teams. Using Mann–Whitney U tests and Wilcoxon signed ranks test, the results of P–S indexes differences among disciplinary teams are summarized in Table 4.18.

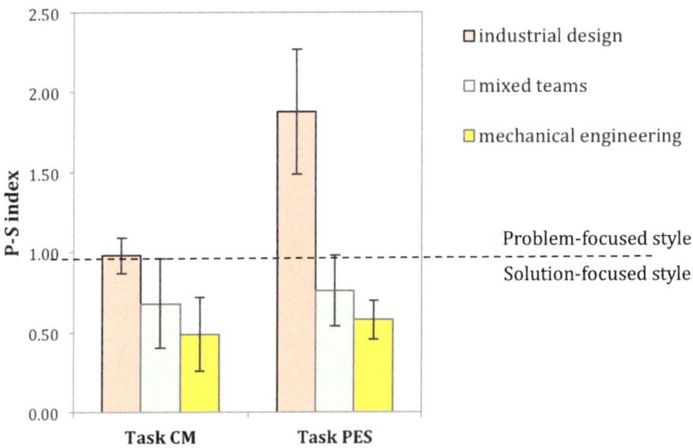

Fig. 4.23 Comparing P–S index of the two halves

Table 4.18 Inter-session comparisons of P–S indexes in the fractioned protocols

Group	First half of design session	Second half of design session
ID teams	ID CM ≈ ID PES	ID CM < ID PES
Mixed teams	Mix CM ≈ Mix PES	Mix CM ≈ Mix PES
ME teams	ME CM ≈ ME PES	ME CM < ME PES
Task CM	ID CM > Mix CM ≈ ME CM	ID CM ≈ Mix CM ≈ ME PES
Task PES	ID PES > Mix PES > ME PES	ID PES > Mix PES ≈ ME PES

≈Not significantly different
>Significantly larger than
<Significantly smaller than

The results suggest that industrial design student teams have a metacognitive design style that is more focused on the design problem than mechanical engineering student teams in both design tasks in the first half of the session. Overall the P–S index dropped in the second half the session.

4.5 Conclusions

This chapter has applied the proposed ontological coding scheme presented in Chap. 3 to multiple cases. Linkography has been used to derive FBS processes. Many coding schemes have been developed for use with design protocols. All such schemes are based on particular views of the activity of designing. Many of these schemes are unique to the data to which they are applied. This limits the applicability of the results obtained. Where more general codings have been attempted they still lack sufficient generality to allow them to be re-used in widely varying circumstances. The use of the FBS ontology and the situated FBS ontology provides a generally applicable coding basis that does not depend on any particular circumstance associated with any unique protocol. The coding scheme does not require that any particular number of designers be involved. It is not limited to any particular stage of the design process, nor is it bound to the face-to-face environment. When it is used to study designers using a computer-mediated environment, especially a 3D environment, the segmentation method needs refinement. However, the principle remains the same.

Figure 4.24 compares the distribution of the FBS processes in the three situations of mechanical engineering students designing using brainstorming, professional architects designing face-to-face and professional architects designing in a 3D-world. The brainstorming session and the face-to-face session exhibit similar patterns of process distributions. The face-to-face session has higher percentages of formulation and reformulation type one, while the brainstorming session has higher

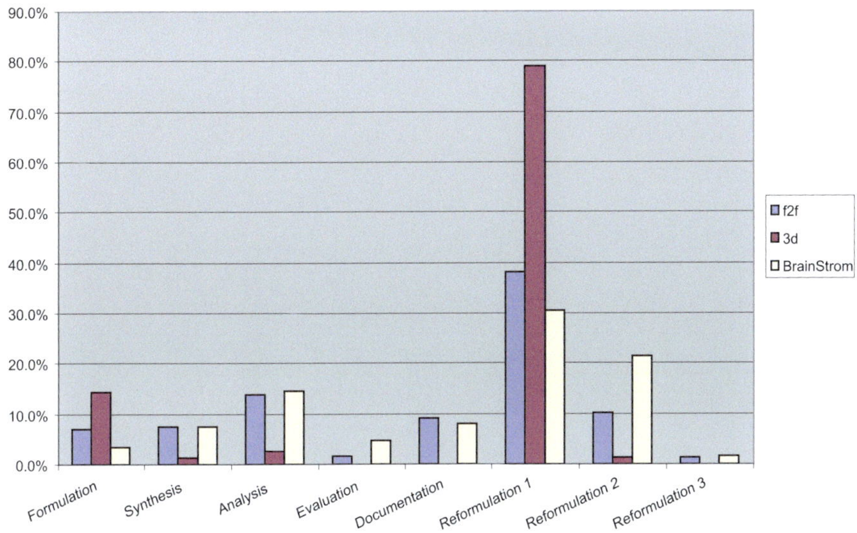

Fig. 4.24 The eight FBS processes of the three sessions

percentages of reformulation type two and evaluation. We can also include the FBS design processes in the 3D-world session, Fig. 4.24.

Figure 4.25 compares the probabilities of each issue in these three sessions. The length (time) for analysis of the face-to-face session is the same as that of the 3D-world session; however, the length of the brainstorming session is four times longer.

From the qualitative analysis we can observe that the designers referred more frequently to the requirements at the beginning of a session. This is reflected by the brainstorming session having the lowest probability of R code. The brainstorming session was more behaviour oriented (higher probability of B codes) while the face-to-face and 3D-world sessions were more structure oriented (higher probability of S codes). The results presented in these cases demonstrate that the ontological approach of coding, together with linkography can be used to study a team of designers and designers using different types of communication channels. These provide a common platform to analyse and compare design activities in various circumstances.

These quantitative models contain considerable detail about the behaviour exhibited in a particular design session. This information can be used to elucidate chronologically-based design behaviour in an individual design session. This can be done by fractioning the design session into halves, thirds, quarters, deciles, etc., treating each fractioned sub-session as a complete session and carrying out the analyses on them. This allows for comparisons of cognitive behaviour across the time.

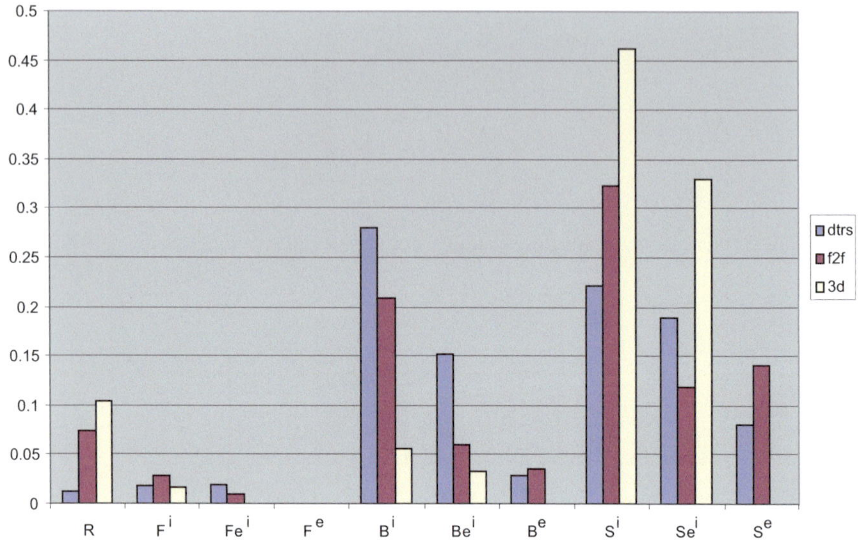

Fig. 4.25 The Markov probabilities of each issue (using the situated FBS coding) of the three sessions

By aggregating results from a set of similar designers, Quantitative models can be used to produce statistically testable results across multiple cohorts. These results can then be compared with aggregated results from design sessions carried out under different conditions to determine the effect of changing design conditions.

Chapter 5
Pilot Study of Statistical and Entropic Description of Linkographs

This chapter presents a pilot study, using the ideas in Chap. 3, to examine two design sessions under two conditions; one being face-to-face and the other using computer mediation to simulate distant collaboration. Preliminary results are presented concerning clustering and entropic measures of the linkograph.

5.1 Two Sessions

This section presents the two cases used in a pilot study, together with the study's qualitative analysis. This data was collected for a larger study concerning team collaboration in a high bandwidth environment (CRC Construction Innovation project, titled: Team Collaboration in High Bandwidth Virtual Environments). Both sessions involved two designers collaborating on an architectural project. In this pilot, only the first sheet of each drawing is studied and compared.

5.1.1 In Situ Face-to-Face Design Session

The first case was an in situ design session carried out in a Sydney architects' office. Two architects, one more senior than the other, were involved in the design of a commercial building in Canberra's city centre. This design session occurred after a review and planning session, subsequent to a client meeting. In this session the designers revisited the relationship between vertical circulation and the void areas, in order to satisfy the client's preferences. The raw data was a video recording of the session. Figure 5.1 is one image from this session and Fig. 5.2 is the first sheet of the drawing that they produced in this session, which will be analysed.

J.W.T. Kan and J.S. Gero, *Quantitative Methods for Studying Design Protocols*, DOI 10.1007/978-94-024-0984-0_5

Fig. 5.1 Face-to-face session; Senior Architect starts drawing the core and the bridges after 4 min

In this session the architects were refining the design after the client's feedback. They needed to add a meeting place on the lobby level; also they needed to work on resolving related issues. The Senior Architect tried to incorporate the client's preferences and requirements, for example, very early in the protocol he mentioned "He wants an atrium", "I mean I know what he wants,", and "… I think he loves the idea of the verge…". During the first 10.5 min of the session, the designers frequently used drawing and gesturing to communicate without explicit verbalising, and nearly all verbalisations were accompanied by non verbal actions; they referred to materials from previous designs; they drew different types of diagrams; sometimes separately; and they referred back and forth to the main plan drawing. Design actions were occurring in parallel, sometimes when the Senior Architect was working on the large drawing the Architect would draw on another sheet of paper or retrieve older drawings. There were interruptions, such as a phone call for about a minute towards the end. The leadership role was clear; the Senior Architect controlled and led the session. The session started with the Architect suggesting a few possible solutions related to the previous session, but the Senior Architect insisted on not jumping to a conclusion and started revisiting the issues and the client's preference by drawing a small plan at the bottom of the sheet (Fig. 5.2). He then

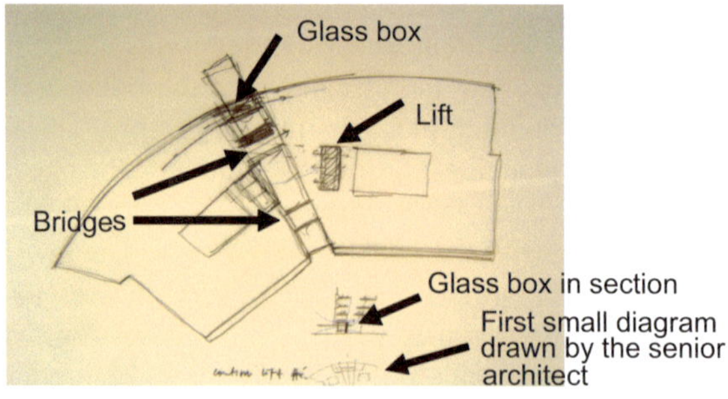

Fig. 5.2 The first sheet of the drawing the architects produced in the first 10.5 min, with annotations added

traced over the position of the bridges, which he regarded as important. Then they discussed the relationship among the lift, void, bridges and lobby. After about 6 min the Senior Architect discovered another problem with the setback of columns. They explored the position of the glass box and its relationship with other levels by drawing a small section with the setback of columns. The designers were dealing more with the structural or formal aspects of the design in this session—where things should be and how they related to each other, in order to satisfy the client. This sheet, Fig. 5.2, was mostly drawn by the Senior Architect; the other Architect drew a small diagram on another sheet.

5.1.2 In Vitro NetMeeting Design Session

The second case was an in vitro session which simulated the distance collaboration of two designers, an Architect and a Landscaper, with the use of computer-mediated tools. Tangible interfaces, Smartboard and Mimio (Smartboard and Mimio are brands of touch-screen interfaces that use different mechanisms. Smartboard is the wall-mounted touch-screen display in Fig. 5.3 and Mimio is attached to a rear-projected horizontal frosted glass table surface), together with Microsoft NetMeeting were used in this experiment. NetMeeting contains a shared whiteboard and a video conferencing tool. A more detailed experimental setup is included in

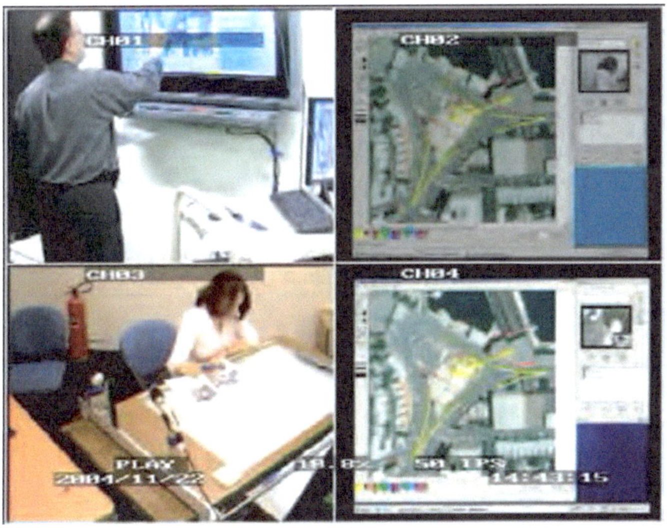

Fig. 5.3 NetMeeting session; the designers translating the issues into drawing at the beginning of the session

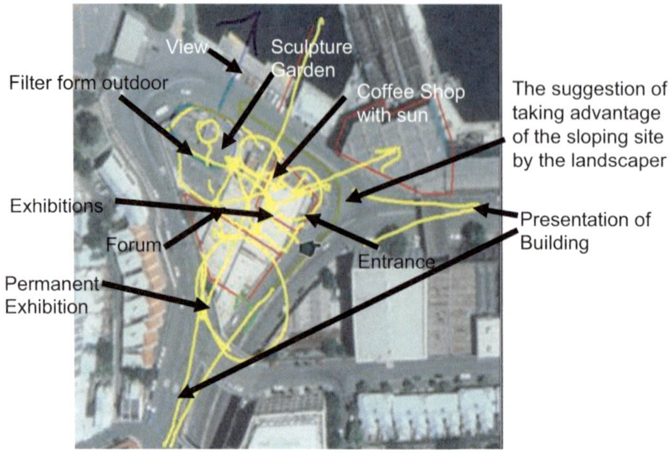

Fig. 5.4 The first sheet of the NetMeeting session, mostly drawn by the Architect

Appendix C. The designers were asked to design an art gallery in a harbour-front triangular site with level changes; the design brief and related materials are attached in Appendix B. Both their displays and actions were recorded as shown in Fig. 5.3. Figure 5.4 is the first sheet that they produced, which will be studied and compared in this pilot study; annotations are added to show the meaning of the drawing.

In this session the designers were given a new design task, so they were focusing more on the functional or conceptual aspect of the design, with time spent on studying the brief. The Architect started the session by trying to figure out the scale of the site in relationship to the brief. He complained that there was nothing there to scale with, and he could only do a mockup. The Landscaper proposed to work out only the appropriate relationship of functional spaces and their approximate sizes. The Architect then started reading the brief aloud with his added interpretation. Following that, the Architect clarified with his partner whether she could see his pointer and started drawing, proposing the main exhibition area on the south side of the site. The Landscaper noticed that there was a level change in the site and suggested taking advantage of that. Reacting to the suggestion, the Architect proposed the location of the central courtyard, entrance and connectivity. Then within the last 1.5 min the Architect produced the remainder of the design in the first sheet with the contribution of the Landscaper, which included the exhibition areas, the coffee shop with northern sun, the sculpture garden with a view, and the forum. Figure 5.4 is the capture of the first page from the screen and the annotations were added by consulting the verbal protocol. Overall we can observe that the Architect took the leadership role in this session and did most of the drawing.

5.1.3 *Qualitative Differences of the Two Sessions*

The design tasks between the two sessions were very different. In the face- to-face session it was a very specific re-design of the circulation space to satisfy the client's requirements for a commercial building. In the NetMeeting session it was about developing a block model preliminary scheme for a proposed art/craft gallery. In the NetMeeting session, they did not have the baggage of previous decisions, whereas in the face-to-face session they had to consider the impact of every move they made on the overall form and structure. For example, they could not just add the requested meeting space to the design without considering the location of the lift.

Although in both sessions the leadership roles were clear, the leadership styles were different. In the face-to-face session, the Senior Architect asserted his leadership by rejecting or correcting ideas suggested by the Architect. For example, after the Architect suggested the location, the Senior Architect responded, "We can't afford the area". In the NetMeeting session the Architect affirmed the Landscaper's idea and developed it further. Some examples of the affirmation are: "that would be gorgeous" and "correct," cause then you…".

The communication in the face-to-face session depended a lot on the use of gestures and tacit knowledge. "… remember we had a central thing here to here (gesturing locations using pen)" is an example from the protocol that contains both tactic knowledge of "a central thing" and gestures of location. In the NetMeeting session, neither assumed the other party had read her/his gesture. The Architect specifically asked "so can you see my mouse? my pointer?" Also, in the NetMeeting session interactions were more sequential and consisted of more affirmations and there was no interaction among gestures. There were more interactions among ideas, drawings, gestures and verbal communications in the face-to-face session.

5.2 Linkography Analysis of Cases

In the face-to-face session, the architects used 10.5 min to finish the first sheet of drawing. Ninety-eight moves were segmented and 299 links were constructed. Of these 98 moves, the architect contributed 38 moves and the Senior Architect contributed 60 moves. In the NetMeeting sessions the Architect and Landscaper took 6.5 min to produce the first sheet, with 97 moves and 277 links. The Landscaper contributed 37 moves and the Architect contributed 60 moves. Figures 5.5 and 5.6 show the linkographs of the two sessions. At the time these linkographs were generated by using a program written in AutoLisp (AutoLisp is a scripting environment within AutoCad). Now there is a publicly available software package to produce the linkograph of an FBS-coded protocol (LINKODER 2011). The critical

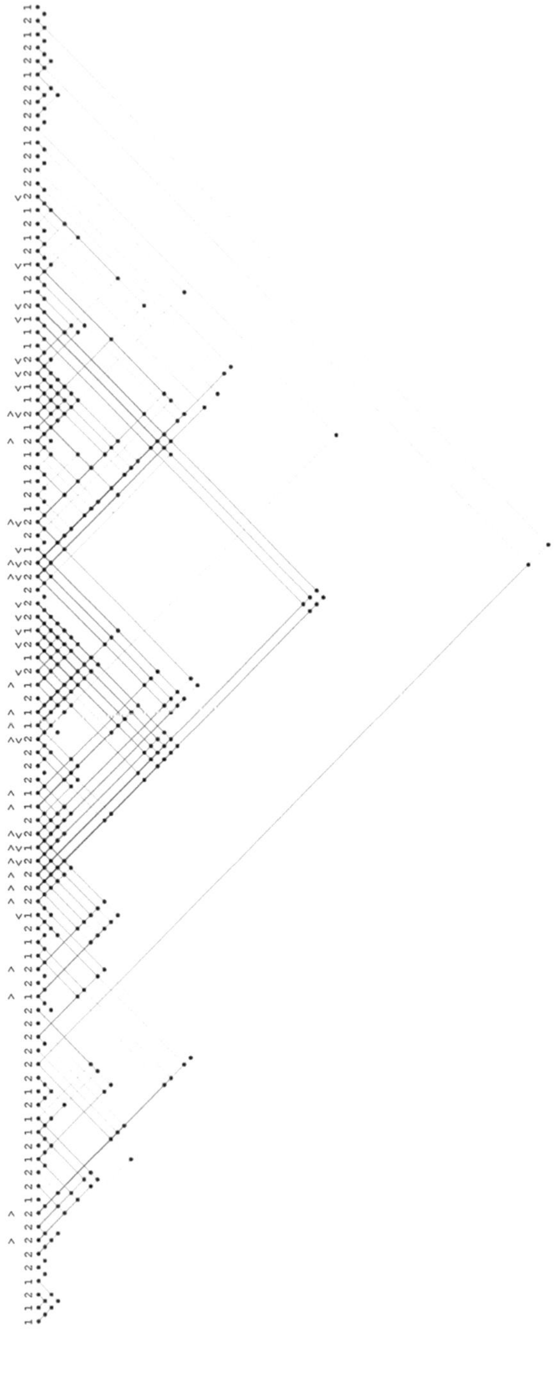

Fig. 5.5 The linkograph of the first 10.5 min of the face-to-face session, with “1” representing the Architect and “2” representing the Senior Architect

Fig. 5.6 The linkograph of the first 6.5 min of the NetMeeting session, with "A" representing the Architect and "L" representing the Landscaper

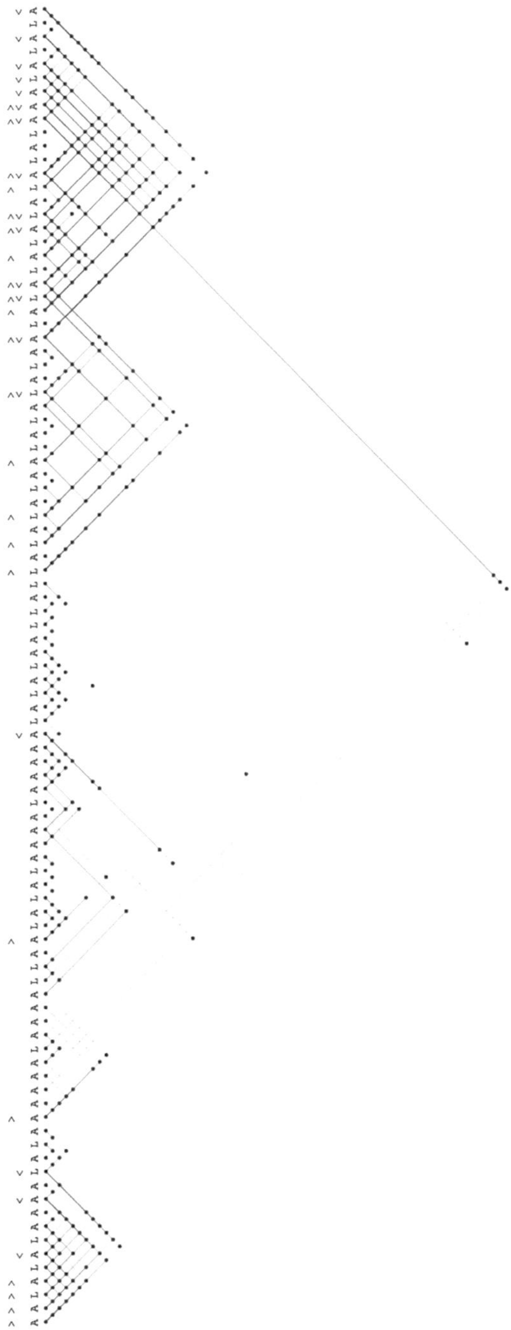

Table 5.1 Critical moves with more than 5, 6, and 7 links of the face-to-face session

	CM^5 (% CM^5)	CM^6 (% CM^6)	CM^7 (% CM^7)
Forelinks	21 (21.4)	17 (17.3)	13 (13.3)
Backlinks	22 (22.4)	15 (15.3)	7 (7.1)
Total	43 (43.9)	32 (32.7)	20 (20.4)

Table 5.2 Critical moves by individuals in the face-to-face session, with A and SA representing the Architect and Senior Architect respectively

	CM^5		CM^6		CM^7	
	A (%)	SA (%)	A (%)	SA (%)	A (%)	SA (%)
Forelinks	7 (18.5)	14 (23.3)	6 (15.8)	11 (18.3)	4 (10.5)	9 (15.0)
Backlinks	7 (18.5)	15 (25.0)	5 (13.2)	10 (16.7)	1 (2.6)	7 (11.7)
Total	14 (36.8)	29 (48.3)	11 (28.9)	21 (35.0)	5 (13.2)	16 (26.7)

The values inside the brackets show the critical move percentages of total moves

moves with more than five links (CM^5) are indicated by: ">" for forelinks and "<" for backlinks. In the face-to-face session, links were dense over the whole session, whereas in the NetMeeting session links were dense towards the end of the session. There was an obvious chunk at the beginning of the NetMeeting session, but not in the face-to-face session.

The link indexes of the face-to-face and the NetMeeting sessions are 3.05 and 2.88 respectively. Tables 5.1 and 5.3 record the critical moves and their percentages over the total number of moves (% CM) of the face-to-face session and the NetMeeting session respectively. The face-to-face session has a total of 43.9 percent of critical moves with more than five links (% CM^5) which is a marginally higher than the NetMeeting session which has a total of 41.2 % CM^5. From these values, the face-to-face session seems to have been more productive than the NetMeeting session. However, the NetMeeting session has a higher % CM^7 than the face-to-face session, 28.9 against 20.4. Tables 5.2 and 5.4 show the breakdown of critical moves by individuals of the two sessions; the critical move percentages of the total number of moves are in brackets. These correspond well with the analysis of the leadership role, with the leaders possessing not only more moves but also higher % CM.

Table 5.3 Critical moves with more than 5, 6, and 7 links of the NetMeeting session

	CM^5 (% CM^5)	CM^6 (% CM^6)	CM^7 (% CM^7)
Forelinks	22 (22.7)	17 (17.5)	14 (14.4)
Backlinks	18 (18.6)	16 (16.5)	14 (14.4)
Total	40 (41.2)	33 (34.0)	28 (28.9)

Table 5.4 Critical moves by individuals in the NetMeeting session, with L and A representing the Landscaper and Architect respectively

	CM^5		CM^6		CM^7	
	L (%)	A (%)	L (%)	A (%)	L (%)	A (%)
Forelinks	8 (21.6)	14 (23.3)	6 (16.2)	11 (18.3)	5 (13.5)	9 (15.0)
Backlinks	5 (13.4)	13 (21.7)	4 (10.8)	12 (20.0)	4 (10.8)	10 (16.7)
Total	13 (35.1)	27 (45.0)	10 (27.0)	23 (38.3)	9 (24.3)	19 (31.7)

The values inside the brackets show the critical move percentages of total moves

5.3 Statistics and Clustering of Links

This section uses the statistical methods described in Chap. 3, Sect. 3.2 to analyse the two sessions.

5.3.1 Statistical Description of the Two Sessions

Tables 5.5 and 5.6 present the statistical descriptions of the linkographs, in reference to the position of links, for the face-to-face and NetMeeting session respectively. Figures 5.7 and 5.8 are the corresponding scatter plots.

The NetMeeting session has a higher X mean and a higher standard deviation than the face-to-face session. This indicates, in general, that the links in the NetMeeting session are distributed more towards the end of the session compared to the face-to-face session. This corresponds to the qualitative analysis of the NetMeeting session, where numerous interrelated actions occurred in the last 1.5 min. The NetMeeting session also has a higher standard deviation, indicating that the nodes are more dispersed than in the face-to-face session. This also matches the qualitative understanding, because at the beginning of the NetMeeting session the designers were trying to figure out how to scale in the shared whiteboard, which formed a separate chunk at the beginning, whereas in the face-to-face session we do not observe this kind of separated chunk. The face-to-face session has a higher mean value of Y suggesting that the face-to-face session has links that are further apart, that is, longer links than in the NetMeeting session. Also, the face-to-face

Table 5.5 Descriptive statistics of the face-to-face session with 299 links

	Minimum	Maximum	Mean	Std. deviation
X location	1.50	97.50	48.23	21.81
Y location	0.50	38.50	3.97	5.10

Table 5.6 Descriptive statistics of the NetMeeting session with 279 links

	Minimum	Maximum	Mean	Std. deviation
X location	1.50	96.50	57.83	29.85
Y location	0.5	34.50	3.60	4.61

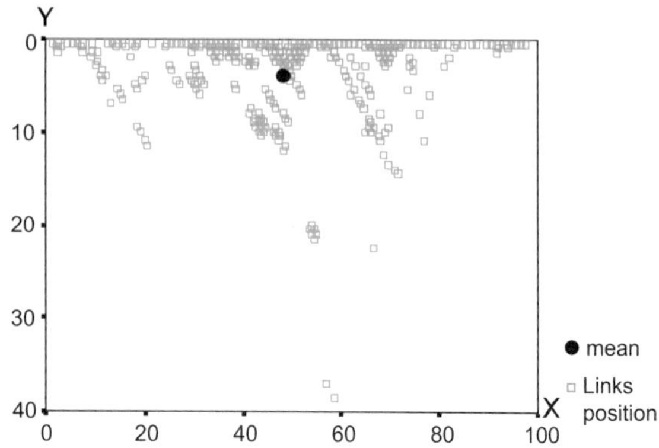

Fig. 5.7 Scatter plot of the face-to-face session

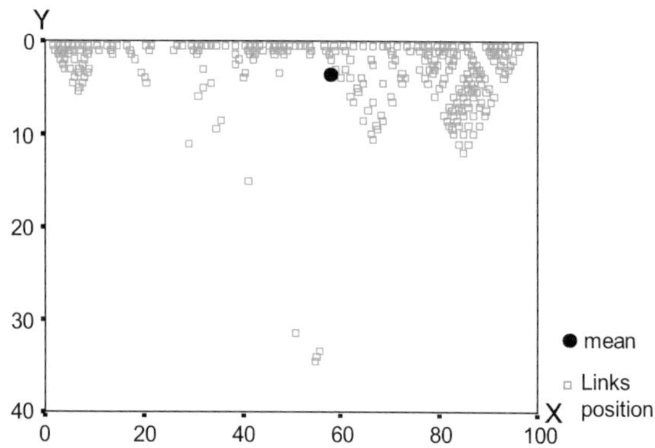

Fig. 5.8 Scatter plot of the NetMeeting session

session has a higher standard deviation, which suggests a greater mixture of long and short links. This agrees with the qualitative analysis, since in the face-to-face session, the designers referred to and traced over their drawing often, causing these long links.

5.3.2 Cluster Analysis of the Two Sessions

Tables 5.7 and 5.8 show the results from the cluster analysis (using SPSS) of the face-to-face and the NetMeeeting sessions respectively; Figs. 5.9 and 5.10 represent their corresponding scatter plots.

Results of the Face-to-Face Session

Figure 5.11 illustrates the variations within clusters and the number of links within each cluster. It shows that the four clusters are well separated, with 95 % confidence that there is no overlapping (in the X direction). It also shows that Cluster 4 is

Table 5.7 Centroids of the face-to-face session

Cluster	X		Y	
	Mean	Std. deviation	Mean	Std. deviation
1	14.18	6.97	2.82	2.95
2	42.50	7.42	3.56	3.29
3	71.67	10.11	3.27	3.72
4	56.39	4.12	24.72	7.43
Combined	48.23	21.81	3.97	5.10

Table 5.8 Centroids of the NetMeeting session

Cluster	X		Y	
	Mean	Std. deviation	Mean	Std. deviation
1	9.04	5.77	1.87	1.43
2	39.40	7.38	2.12	2.97
3	85.61	5.06	4.09	3.00
4	65.13	5.65	3.51	2.98
5	53.88	2.29	33.38	1.31
Combined	57.83	29.85	3.60	4.61

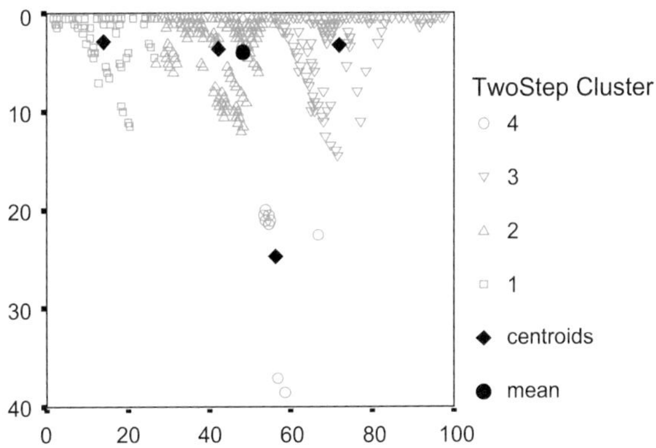

Fig. 5.9 Scatter plot of the clusters of the face-to-face session

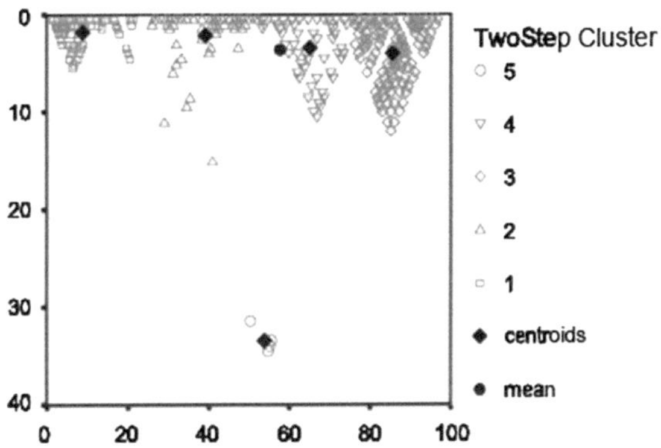

Fig. 5.10 Scatter plot of the clusters of the NetMeeting session

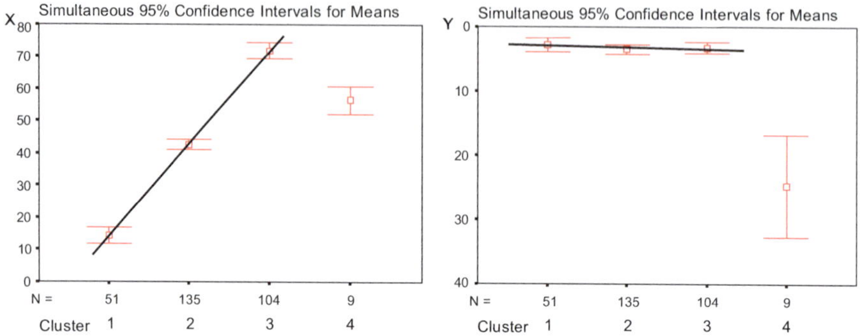

Fig. 5.11 Within cluster variation for the face-to-face session

different from the others. Cluster 4 seems to be the outlier in statistical terms; it contains only nine links. Neglecting this cluster for this moment, Clusters 1, 2 and 3 map well with the qualitative analysis. In Cluster 1, with 51 links, the two designers were discussing issues arising from the previous meeting. In Cluster 2, the biggest cluster, with 135 links, the Senior Architect started drawing and they were considering the behavioural impact of moving the lift, void and bridge. In Cluster 3, with 104 links, the Senior Architect realised another issue induced by the setting-back of columns. Examining Fig. 5.9, Cluster 4, the statistical outlier, contains all the long links; it groups those links that are far apart which link Cluster 3 with Cluster 2 and Cluster 1. In this particular case, these links were formed either because the participants were tracing over or referring to depictions that they drew earlier or when they were concerned with the symmetrical axis of the building.

Results of the NetMeeting Session

Figure 5.12 shows the variations within clusters and the number of links within each cluster. It demonstrates that the five clusters are well separated, with 95 % confidence that there is no overlapping. The statistical outlier, Cluster 5, contains

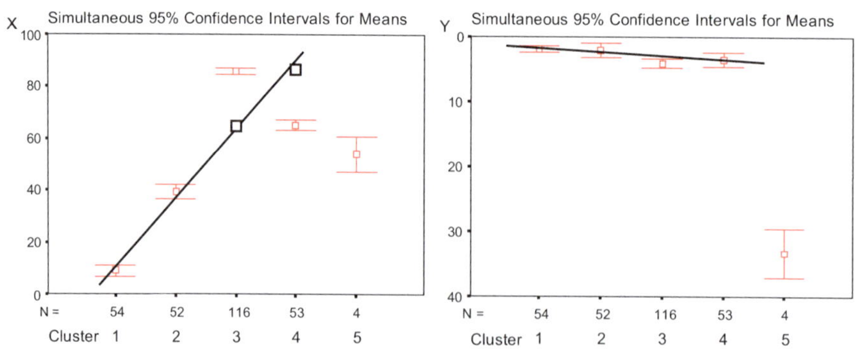

Fig. 5.12 Within cluster variation for the NetMeeting session. Squares were added to represent the clusters in the correct time sequence

only four links. Again, neglecting it, the other clusters reflect the themes of the protocol. Readers should note that the cluster numbers here do not reflect the time sequence; the time sequence would be Cluster 1, Cluster 2, Cluster 4 and then Cluster 3. This is represented by adding two squares in the X mean in Fig. 5.12. This happened because the number of clusters and their labels were automatically generated by the SPSS software. The labels did not carry any notion of time sequence; the software calculated the distances among nodes, grouped and labeled them into clusters without the foreknowledge of how user will interpret the results.

In Cluster 1, with 54 links, they were discussing constraints imposed by NetMeeting—how to scale without references. There were 52 links in Cluster 2; this began with the Architect reading from the brief and continued with the concerns regarding the functional spaces and their relationship. In Cluster 4, with 53 links, the Landscaper introduced another idea, suggesting they take advantage of the level changes in the site, which led to further development by the Architect in Cluster 3, the biggest cluster, 116 links. In this major cluster, the Architect proposed the location of the majority of functional spaces in relation to the site and each other. Cluster 5, the statistical outlier, contains four links between Cluster 5 and the functions in Cluster 2. The Architect was referring to his interpretation of the brief when proposing the location of these functional spaces. The mean value of Y slightly increases towards the end of the session, indicating that links are getting longer.

Comparing the Results of the Two Sessions

Although the face-to-face session has more links, it contains one fewer cluster than the NetMeeting session. This is due to the content; the extra cluster in the NetMeeting session is more about commenting on the technology than designing. So both sessions contain three groups of ideas, but the distribution of the number of links is quite different. The NetMeeting session contains one major cluster at the end with twice the number of links than other clusters. In the face-to-face session, there are two large clusters and one small cluster, and the larger one is in the middle. The statistical outliers contain all the long links; the face-to-face session has a bigger cluster of long links than the NetMeeting session.

The clusters, except the statistical outliers, indicate idea chunks. The size of a cluster indicates its relative importance and the number of clusters suggests the amount of ideas within the period studied. Both sessions have an equal number of ideas chunks but the NetMeeting had one important idea at the end of the first sheet of drawing. The distance (Distance between centroids are usually calculated by $\sqrt{(\overline{x}_i - \overline{x}_j)^2 + (\overline{y}_i - \overline{y}_j)^2}$, where $(\overline{x}_i, \overline{y}_i)$ and $(\overline{x}_j, \overline{y}_j)$ are the centroids. However, here the distance is computed by $(\overline{x}_i - \overline{x}_j)$, because the interest is in relating it to the distance between moves, and also the $\overline{y}$ variation is comparatively small enough to be insignificant.) between centroids of subsequence clusters is shorter in the NetMeeting session. In the face-to-face session the distance between Cluster 1 and 2 is 28.3, and the distance between Cluster 2 and 3 is 29.2. So the average is 28.7. In the NetMeeting session the distance between Cluster 1 and 2 is 30.4, the distance between Cluster 2 and 4 is 25.7, and the distance between Cluster 4 and 3 is 20.5. There is a steady decrease of distance between clusters and the average distance is 25.5. This is not obvious when looking at the design protocols or watching the video. In the face-to-face session, there were interruptions at the beginning and at the end, which slowed the pace, and the distances between the three clusters were even. In the NetMeeting session, the designers were moving more quickly between idea chunks; this is shown by the shortening of subsequent inter-cluster distances. More quickly here means in terms of protocol segments not in terms of time.

Clusters and the distances between them are the basis for new insights in the qualitative understanding of a design session and through multiple design sessions into designing.

5.4 Entropic Measurements

This section presents the results of the entropic measurement of the e-to-face and NetMeeting sessions and the entropic measurement of individuals. It also explores the trends of entropic variation across a design session. How to measure the entropy of a linkograph is described in Chap. 3, Sect. 3.3.2.

5.4.1 Entropy of the Two Sessions

Tables 5.9 and 5.10 show the entropy of the face-to-face and NetMeeting sessions respectively. Forelinks can be seen as initiations and backlinks as responses. As proposed earlier, a higher value of the entropy (labelled H) of forelinks signifies greater opportunity to initiate design moves, and a higher H value of backlinks denotes greater opportunity to build upon previous design moves. The horizonlink entropy indicates the opportunity according to the length of the links; high values of entropy usually indicate a mixture of long and short links, which suggests the cohesiveness and incubativeness of ideas. In the face-to-face session the backlink entropy is slightly higher than the forelink entropy, which indicates a greater opportunity of building upon rather than initiating moves. The NetMeeting session scored the opposite in terms of its entropy measurements, which indicate the initiation opportunity was greater than the response opportunity. These results tentatively match our qualitative analyses of both sessions. In the face-to-face session the designers were at the stage of refining the design, referring to what is already there, whereas in the NetMeeting session they started from the beginning, initiating new

Table 5.9 Entropy of the face-to-face session

Forelinks total H	Backlinks total H	Horizonlinks total H	Cumulative total
34.171	36.693	12.244	83.109

Table 5.10 Entropy of the NetMeeting session

Forelinks total H	Backlinks total H	Horizonlinks total H	Cumulative total
27.865	26.922	11.477	66.264

ideas. However, the difference in entropies is too small to be conclusive. Both sessions have similar horizonlink entropy. Overall, the face-to-face session has higher entropy in all three areas, implying the opportunities are greater in all areas.

These results concur with the link index study; the link indices of the face-to-face and the NetMeeting sessions are 3.05 and 2.88 respectively (Sect. 5.2). However, the cumulative entropy measure shows a larger percentage difference than the link index study (25.4 vs. 5.9 %). This may help to discern subtle differences. Also the link index is a measure of saturation and does not separate the contribution by forelinks or backlinks while entropy measures each separately.

5.4.2 Entropic Measurement of Individuals

Chapter 3, Sect. 3.3.2 has described how the entropy of each move is measured for the forelink and backlink. If all the moves contributed by an individual are singled out, the forelink and backlink entropy contributed by that individual can then be calculated. The entropy of individuals is measured to see if it matches their observed role and participation. Tables 5.11 and 5.12 are the forelink and backlink entropy contributions of the various participants. In both sessions the leaders scored higher than their partners in both forelink and backlink entropy. There are two factors that contribute to this: the number of moves and the entropy per move. From our qualitative analysis we know the leaders did most of the drawing, hence they contributed more moves. The leaders also have a higher entropy per move, except for the forelinks of the Landscaper. This is due to the Landscaper's contribution of a new idea—taking advantage of level changes, which is an opportunistic initiation.

Table 5.11 Forelink and backlink entropy by the senior architect and the architect in the face-to-face session

	Moves	Forelink H		Backlink H	
Senior architect	60	21.661	0.361 per move	22.846	0.381 per move
Architect	38	12.511	0.329 per move	13.847	0.364 per move

Table 5.12 Forelink and backlink entropy by the architect and the landscaper in the NetMeeting session

	Moves	Forelink H		Backlink H	
Architect	60	16.582	0.276 per move	17.930	0.299 per move
Landscaper	37	11.283	0.305 per move	8.988	0.243 per move

The individuals' entropy scores reflect their opportunistic contributions. This contrasts with the CM study in Tables 5.2 and 5.4, which indicate that the leaders in both sessions have higher % CM in both forelinks and backlinks.

5.4.3 Changes in Entropy During the Session

Observing the linkographs of Figs. 5.5 and 5.6 it can be inferred that the entropy varies across the time line. There are at least two possible ways to measure this change; one uses a fixed time frame as a reference window and the other uses a fixed number of moves as the width of the window to calculate the moving average of the entropy. The latter is used because it is easier to operate and provides a more meaningful comparison. For example, entropy can be calculated within a seven-move window, as in Fig. 5.13. The calculation starts from the first move and advances to the next move, until the window reaches the end. The changes of entropy across the design session can then be recorded. Those links outside the window, not inside the shaded triangle, are disregarded.

The seven-move cut is indicative rather than conclusive. With a large linkograph, using a seven-move window will ignore too many links that make the analysis insignificant. A suitable window width for obtaining meaningful results will be derived theoretically and empirically. By monitoring the change in the entropy, the trend of a design session can be studied and compared.

Determining the Width of Moves Window

The seven-move window is inspired by Miller's magic number seven (Miller 1956). He demonstrated that the chunk of information held in the short-term memory was limited to seven plus or minus two. There are other more articulated memory models, like Paivio's (1986), Bartlett's (1932), and Logie's (1995, 2001), which use the term "working memory" rather than "short-term memory". Baddeley's model of working memory contains three parts: the visuo-spatial sketch pad, the phonological loop, and the central executive. Logie developed Baddeley's model to consider knowledge, long-term memory representation, as a filter that will bias perceptions before getting into the three parts of working memory. It is considered that the content of this memory degrades rapidly; in general it is believed it holds

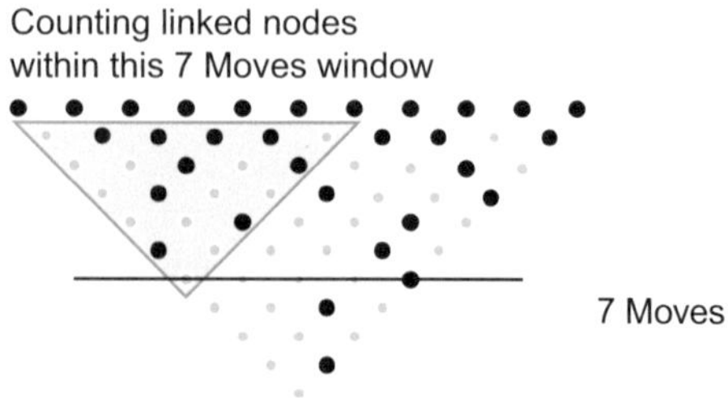

Fig. 5.13 Illustration of using a seven-move window to calculate the entropies

information for about 12–20 s. Important or interesting information will be sustained in the working memory and will trigger further associations in the memory system. As we assume that moves are a selected externalisation of the designers' cognitive processes, in order to communicate with their partners, the cognitive processes that correspond to the moves can be in the working memory or in long-term memory. When these processes are in the working memory, the corresponding moves will have high interconnectivity.

Experimenting with a seven-move window or a nine-move window showed that they were not capturing enough links and the graphs did not reveal any trend. The graphs smoothened as the window width increased. When the moves window was widened to 28 moves, Figs. 5.14 and 5.15 were obtained. For ease of comparison, the value of the entropy was normalised for each window, by dividing it by the window width, 28, to obtain the average entropy per move. With this window width all the disregarded links are those inside the statistical "outliers", as described in Sect. 5.3.2. Also, this window width is slightly wider than the average inter-cluster distance.

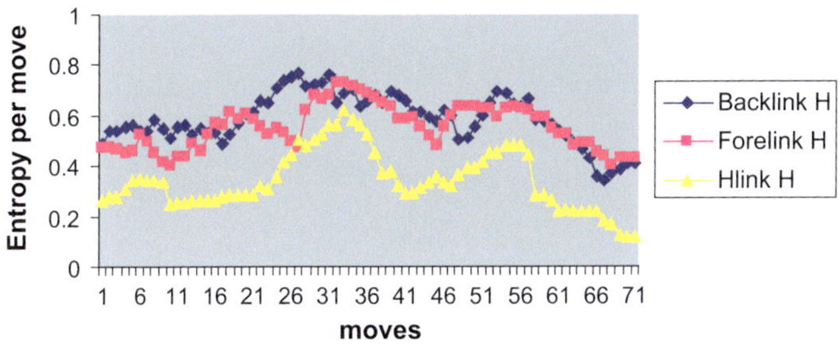

Fig. 5.14 Entropy variations in the face-to-face session, using a 28-move window to calculate entropy

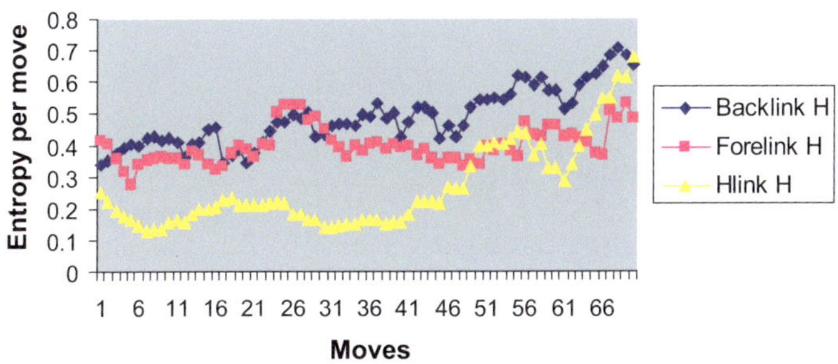

Fig. 5.15 Entropy variations in the NetMeeting session, using a 28-move window to calculate entropy

Observations of Entropy Changes

The trends of the three different types of entropy look similar in their respective sessions but in general the horizonlink entropy scores lower. The overall trend of the two sessions is quite different. In the face-to-face session the entropies peak in the middle of the session, while in the NetMeeting session the entropies peak at the end of the session. This complements the cluster analysis; the clusters that have the most links receive the highest entropy.

To further investigate the trends the backlink entropies were selected and fitted by a polynomial function. They were assigned to an array in MatLab and its supervised polynomial fit function was used to obtain a fourth-degree polynomial, Figs. 5.16 and 5.17. Since the scale of the entropy axis of Figs. 5.16 and 5.17 was

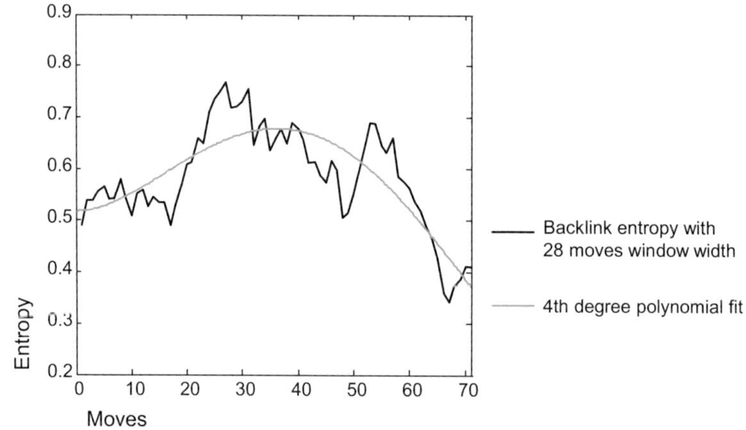

Fig. 5.16 Polynomial fit of the entropy variations in the face-to-face session

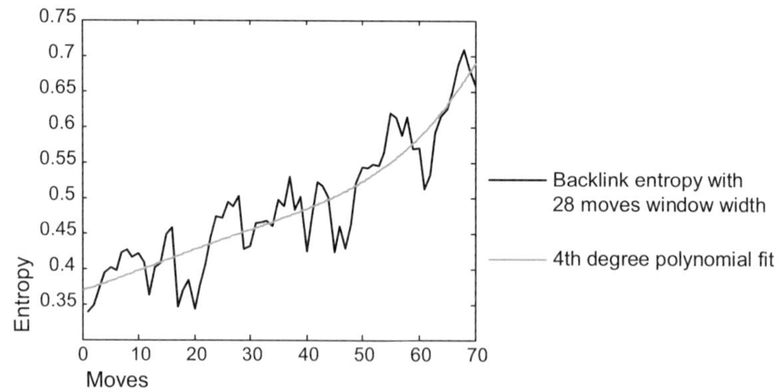

Fig. 5.17 Polynomial fit of the entropy variations in the NetMeeting session

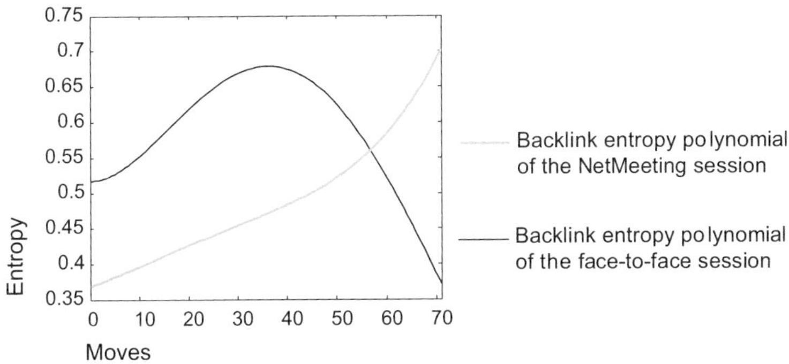

Fig. 5.18 Plotting the backlink entropy polynomial fit of the two sessions with the same scale

different, the polynomials were re-plotted with the same scale in Fig. 5.18 for comparison.

The form of the two polynomials is very different. The rate of entropy variation (the slope of the curve) of the NetMeeting session is always positive, while in the second half of the face-to-face session the rate of entropy variation is negative. Figure 5.19 plots the rate of the change in entropy of the two sessions. The face-to-face session has a higher backlink entropy, while the NetMeeting session has a higher positive rate of change in entropy. This is confirmed by using adaptive Simpson quadrature in MatLab to calculate the areas under the curves in Fig. 5.19. The areas are 0.032 for the face-to-face session and 0.162 for the NetMeeting session. This can be seen as the signature of a design session.

This entropy variation rate can be correlated to the idea chunks distance in Sect. 5.3.2. A positive rate means the ideas chunks are getting closer and closer as the design move along and a negative rate means the ideas chunks are getting further apart as time goes by.

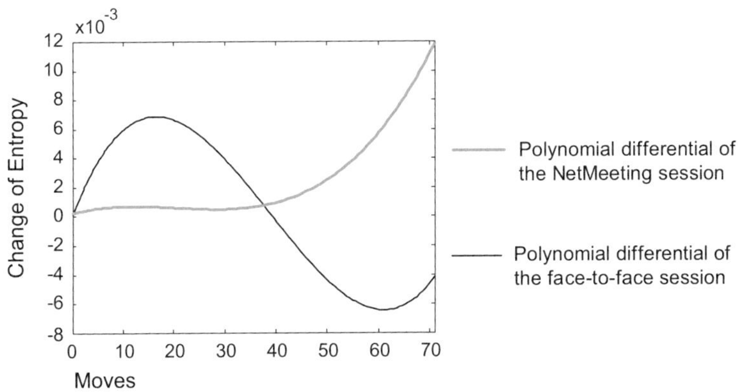

Fig. 5.19 The polynomial representing the change of the backlink entropy in both sessions

5.5 Findings and Discussions

In this study the proposed methods of studying design protocol were applied to two design sessions. In both sessions the designers were working together on the same artefacts simultaneously in two different media. In the face-to-face session they used traditional paper and pencil, whereas in the NetMeeting session they used a computer-mediated simulation of distance collaboration by a shared whiteboard and video-conferencing tool. There were more interactions between the designers, with gestures and reference to drawings, in the face-to-face session. Many of the verbal communications were not complete, because of the tacit knowledge. In the NetMeeting session the turn-taking in conversation was more orderly and complete.

Standard descriptive statistics were able to describe the shape of a linkographs and in our case study they were able to pick up some of the differences in the design processes, such as the lengths of the links and the position of intensive activities. The preliminary results using clustering and entropy were promising. Clusters automatically generated by commercial software were able to map onto the actual design activities, hence the semantics of a cluster could be labelled. The statistic outliers contained long links that connected other clusters. Entropic measurement matched the qualitative analysis. Traditional studies of linkographs have used link index and critical moves to analyse the design protocol. In the two case studies, the total cumulative entropies agreed with the link index with a different magnitude.

From the investigation of entropy variation, it can be observed that the two sessions produced very different shapes of move-entropy graphs. Further investigation is required for discerning the meaning of this signature. It is likely that a positive rate of change of the move-entropy graphs implies a diverging process, since the idea development opportunity is continually increasing. It is likely that a negative rate of change of the move-entropy graphs implies a converging process, since the idea development opportunity is continually lessening. We explore this further in the next chapter.

Chapter 6
Entropic Measurement and Design Outcome

This chapter presents an application of the entropic measurement of linkographs. Linkographs from the empirical protocol studies of six architects are used to verify whether the design outcome correlates with some entropy measure of the linkograph.

6.1 The Experiment

Again, Bilda's experimental data and link data (Bilda 2006) were used in this chapter. The aim of that research was to understand the role of imagery and sketching in the conceptual phase of the individual design process. Experiments with designing in a blindfolded condition (participants were not allowed to sketch during designing, only at the end of session) were set up for comparison with designing under normal (participants allowed to sketch) conditions. This chapter examines whether the differences in the design outcomes can be reflected in the entropic measures of their corresponding linkographs. Below is a summary of the experimental setup.

The six architects who participated (two females and four males) have each been practising for more than 15 years. Architects A1 and A2 run their own companies and have been awarded prizes for their designs in Australia; Architect A3 is a senior designer in a well-known architectural firm. These three participants were teaching part-time in design studios. A4 works for one of Australia's largest architectural companies and has been the leader of many residential building projects, from small to large scale. A5 is one of the founders and directors of an award-winning architectural company. A6 is a very famous residential architect in Sydney, and he directs his company, known by his name, with 50 employees. They all had experience with designing residential buildings.

Each architect was asked to participate in two design sessions; each lasted about 45 min, with at least 1-month's separation between them. They were asked to talk

J.W.T. Kan and J.S. Gero, *Quantitative Methods for Studying Design Protocols*, DOI 10.1007/978-94-024-0984-0_6

(a)

(b)

Fig. 6.1 **a** Blindfolded session followed by quick sketching, **b** sketching session (Bilda 2006)

aloud while designing and were tutored in doing so. In one session the designer was blindfolded (BF) during designing, so that s/he was unable to sketch, and then s/he was asked to quickly draw what s/he had designed at the end of the session with the blindfold taken off, as depicted in Fig. 6.1a. At least 1 month later, the same architect designed in their normal mode, that is, using sketching (SK), Fig. 6.1b. The same site was used, so the 1-month separation helped to minimise the possibility of familiarisation with the site if the sessions had been held closer together. Bilda (2006) did not explain the time separation in his thesis. Both sessions involved designing a house with different requirements, brief 01 and brief 02; site plan and site photos were provided. Design brief 01 required them to design a house for two artists: a painter and a dancer. The house was to have two studios, an observatory, a sculpture garden and living, eating, sleeping areas. Design brief 02 asked them to design a house, on the same site as design brief 01, for a couple with five children aged from 3 to 17, that would accommodate children's and parent's sleeping areas, family space, study, guest house, eating and outdoor playing spaces.

The sessions were video- and audio-recorded as raw protocol data. The participants were organised into two groups of three; the first group started with the BF session with brief 02, followed by the SK session with brief 01. In the second group the sequence was reversed and the briefs were swapped. This was undertaken to ensure the design outcomes were not biased by the brief or the sequence. Bilda (2006) did not explain the reversing of the sequences and the swapping of brief. The details of the experimental procedure can be found in Bilda (2006) or Bilda et al. (2006).

6.2 Design Outcome

In order to minimise subjectivity in the assessment, the design outcomes, sketches from the 12 design sessions, were double-blind reviewed by three judges (Amabile 1996). All the judges have practised and taught design for more than 15 years. They

Table 6.1 Cumulative score of the criteria in design sessions by three judges

	Innovative	Creative	Brief	Practical	Flexible	Average
BF01	4.0	5.3	7.7	7.7	6.0	6.1
BF02	4.3	6.0	6.3	7.0	6.3	6.0
BF03	6.0	6.3	7.7	7.0	7.3	6.9
BF04	5.0	5.7	7.5	6.7	5.7	6.1
BF05	6.3	7.3	8.0	7.7	6.0	7.1
BF06	4.3	3.7	5.7	5.0	5.7	4.9
SK01	4.3	5.0	6.3	6.0	5.3	5.4
SK02	5.3	5.7	6.3	5.7	6.3	5.9
SK03	6.7	7.3	6.3	5.3	6.7	6.5
SK04	4.3	4.7	5.0	3.7	4.0	4.3
SK05	6.0	6.3	7.0	7.0	5.7	6.4
SK06	4.0	4.7	5.3	5.7	5.0	4.9

were unaware of the experiment. The judges were given the design briefs and other materials. They were asked to assess the photocopies of the sketches according to five categories: innovation, creativity, satisfaction of the design brief, practicality and flexibility. A 10-point scale was used. Bilda (2006), using Kendall's coefficient testing, considered that the concordance between the three judges was sufficiently high to accept their scores as valid measures of the design outcomes.

Table 6.1 summarises the average score of the assessment criteria of the 12 design sessions, where the number following SK or BF is the designator of the particular designer. Examples of design outcomes are shown in Figs. 6.2 and 6.3. The following were observed:

1. the average score of the BF sessions was higher than the SK sessions;
2. all the BF sessions received higher scores than the corresponding SK sessions, with one exception that had an equal score (SK06 and BF06), the score difference ranged from 0.1 to 1.8;

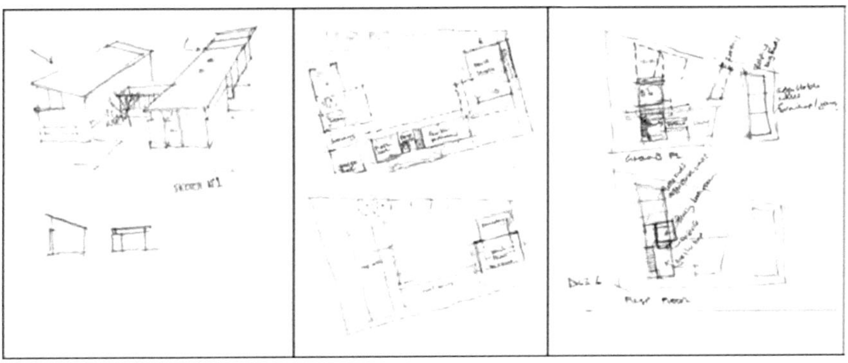

Fig. 6.2 Examples from high-scoring sessions, from left to right BF05, BF03, and SK03

Fig. 6.3 Examples from low-scoring sessions, from left to right SK04, SK06, and BF06

3. the highest average score was 7.1 (BF05) and the lowest score was 4.9 (SK06);
4. adding the SK and BF scores showed that Architect 05 received the highest score and Architect 06 received the lowest score;
5. the scores for individual categories ranged from 3.7 to 8. There were two 3.7, the lowest, and three 4 s scattered in different categories. There were one 8 and four 7.7 s clustered around the blindfold sessions in the practical and satisfying the design brief categories.

Performing statistical testing (paired-*t* test) on observations 1 and 2 revealed that the score differences between the SK and BF sessions were not significant ($p = 0.068 > 0.05$).

6.3 High- and Low-Scoring Sessions

When the three highest-scoring and the three lowest-scoring sessions were grouped together, the score difference between them was over 40 %, which performing *t* test with unequal variance was found to be statistically significant ($p = 0.0014 < 0.05$). The innovative and creative categories were the main contributors to this score difference. The highest-scoring sessions were BF05, BF03 and SK03 and the lowest-scoring sessions were SK04, SK06 and BF06. Figures 6.2 and 6.3 show the sketches of the high and low scored sessions respectively. SK04, the lowest-scoring session, was considered the least "practical" and least "flexible" design solution. It also scored lowest in terms of fulfilling the design brief. BF06 was judged as the least "creative" design solution. At the other end, the high-scoring sessions, BF05 and SK03 were considered to be the most creative. BF05 scored highest in terms of fulfilling the design brief, and shared with one other session in being considered the most "practical" design.

6.3.1 *Qualitative Comparison of the Highest- and Lowest-Scoring Sessions*

In the lowest-scoring session, SK04, the designer started the session by analysing the site and considering the environment. Very early in the session the designer decided to have a two-level building (about 1 min 15 s into the session) near the south boundary. The designer then started laying out the spaces and writing down the requirements. On the ground floor, garage, living area, kitchen, bathroom, master bedroom, and laundry were considered in sequence. The location of the stairs was then positioned at around 9 min. The spaces were repositioned before considering the upstairs plan. Then the locations of bedrooms on the upper level were decided and a gallery link between them was proposed. The designer was not happy with the facade and decided "to reduce the amount of bulk on the top floor" (at around 18 min). The designer then checked whether all the spaces in the brief were being covered. In the remaining 28 min the designer mainly focused on relocating those spaces, changing the size and proportion of those spaces, and adding details. Some examples of verbal protocol are: "reduce those first two rooms slightly and enlarge the last one", "they were in the old positions and putting in the new positions", "should change proportions now". In this session, the designer spent a lot of time in the solution space, time was spent in solving problems created by having two storeys. The main idea seemed to be the central gallery, but it was not obvious in the drawings. Other ideas like privacy, light and views were considered but the final sketches bore no evidence of these.

In the highest-scoring session, BF05, the designer started commenting and analysing the requirements of the studio spaces. There were ideas of separating the dance studio from the painting studio, a courtyard between the studios, a long line of studio space, southern light for the painting studio, northern light versus southern light, borrowing a design from Glasgow Art School studio's ceiling with "big banks of south-facing light", and celebrating westerly light. Stacking up bedrooms or keeping them all single-storey were also considered. Afterwards, connections from the living area to the site were proposed. An "H" plan with two big "C" sections were considered. An open connection without roof was proposed. The following verbal protocol suggested, at around 19 min, that the designer had solidified those concepts and had a rough idea about the form: "So there's two forms, the big linear form of the dance studio spaces…". After that the designer worked on those details, such as "I am going to put a high wall on the southern side", decided how to enter the building and the levels of different spaces. Towards the end of the session a mental walk-through of the building was performed. This involved, for example, visualising the ceiling line and suggesting the location of the gutter. Finally, it was decided that the observatory should be in the courtyard, an outdoor space. The final sketches contained those design ideas that were observed in the verbal report, such as the "H" plan, the linear studios, the south-facing light, and the open connection.

6.4 Constructing Linkographs

According to Bilda (2006), the verbal protocols were segmented by inspecting the designer's intention, similar to the approach used by Suwa et al. (1998). Table 6.2 shows an excerpt from one of the protocols. The average segment length of the twelve sessions ranged from 17 to 26 s.

The links were constructed with the aid of searching for keywords and searching those keywords to find the segment. In the SK sessions, video footage was consulted, while in the BF sessions only verbal protocol was used to discern the links. Twelve linkographs were produced. Figures 6.4 and 6.5 represent the linkographs of the highest-scoring session and the lowest-scoring session respectively. Linkographs of all the sessions were documented in the Appendix D.

Table 6.2 An example segmentation of protocol

Time	Segment number	Segment content
15.50	51	Look, the thing that I'm thinking now is that because I've got such an overwhelming desire to design a courtyard house, and I think that in this kind of situation where you've got a very large site and, umm, a semi-public space that it can borrow, in a way, (16.07) that what you'd start to plumb for is a courtyard building; parts of which are built and parts of which are unbuilt
16.14	52	So, I'd be inclined to organise the dancer's studio and the living spaces and parts of the, the bedrooms… (16.29) or no, the bedroom spaces I think should go down to the east… to give them some separation…
16.34	53	So I'm imagining now a broken form, something that's got the courtyard essentially as its organising structure, but which then has parts built, parts unbuilt

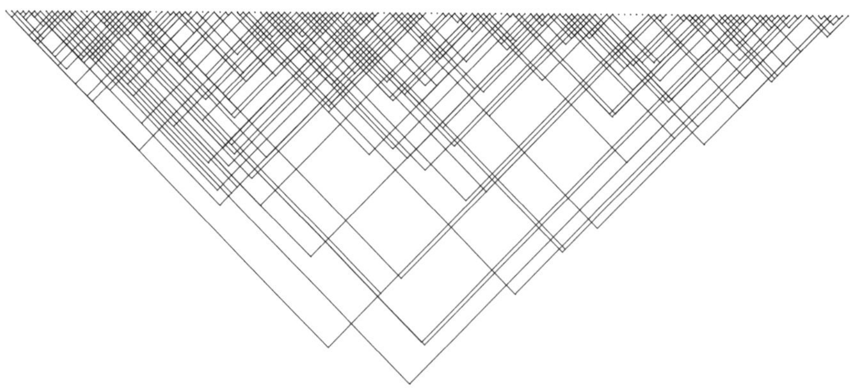

Fig. 6.4 The highest-scoring sessions

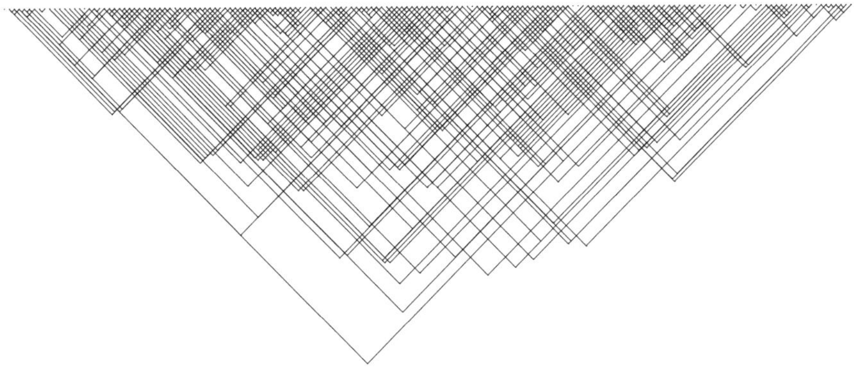

Fig. 6.5 The lowest-scoring sessions

6.5 Entropic Measurement

Table 6.3 shows the normalised entropy of each sessions together with their overall score. For ease of comparison, the normalised values were used: linkograph entropy divided by the number of moves, instead of absolute values.

Overall the BF sessions had higher entropy than their corresponding SK session, with one exception. The differences in entropy were marginal and the evidence was insufficient to suggest BF sessions had higher entropy than their corresponding SK session ($p = 0.14 > 0.05$ with paired t test). Positive correlation between entropy and the evaluation of the design outcome was weak. Some of the lowly ranked design sessions had high entropy values. The correlation coefficient r of the session's entropy and the corresponding outcome was −0.35. Therefore any hypothesis that the design outcome directly correlates with the entropy of the linkograph is not supported by the data from this study.

Table 6.3 Entropies of the 12 sessions

	nBH	nFH	nHH	Total	Outcome
BF01	0.125	0.122	0.060	0.307	6.1
BF02	0.161	0.155	0.066	0.383	6.0
BF03	0.143	0.140	0.055	0.338	6.9
BF04	0.240	0.220	0.093	0.553	6.1
BF05	0.224	0.193	0.082	0.499	7.1
BF06	0.188	0.189	0.105	0.481	4.9
SK01	0.137	0.124	0.077	0.337	5.4
SK02	0.157	0.150	0.065	0.373	5.9
SK03	0.124	0.131	0.044	0.299	6.5
SK04	0.227	0.203	0.098	0.529	4.3
SK05	0.176	0.125	0.071	0.372	6.4
SK06	0.184	0.175	0.063	0.422	4.9

nBH: normalised backlink entropy
nFH: normalised forelink entropy nHH: normalised horizonlink entropy

Table 6.4 Entropies of the high- and low-scoring session

	Session	Score	Entropy per segment
High score	BF05	7.1	0.499
	BF03	6.9	0.338
	SK03	6.5	0.299
Low score	SK04	4.3	0.529
	BF06	4.9	0.481
	SK06	4.9	0.422

6.5.1 Entropy of High- and Low-Scoring Sessions

Table 6.4 shows the scores and normalised total entropies of the three high- scoring and the three low-scoring outcomes of sessions in the descending order of entropy in its own category. The entropies of the high-scoring sessions correlated with the design outcome ($r = 0.86$) but the low-scoring sessions negatively correlated with the outcome ($r = -0.83$). The correlation coefficient is for reference only because the sample size is too small to be conclusive. The low-scoring sessions have higher entropies than the high-scoring sessions, which indicate more links among moves. This suggested that more links in the design process does not necessarily produce better designs. The design outcomes have been shown in Figs. 6.2 and 6.3.

6.6 Entropic Variations

In Sect. 5.4.3 it was suggested that the entropic variation and the rate of variation in entropy can be seen as the signature of a design session. In this section the three highest-scoring and the three lowest-scoring sessions were selected to be studied, because their score differences gave grounds for comparison.

6.6.1 Entropic Variation Graphs

Using a 28-move window, as described in Sect. 5.4.3, the changes in entropy graphs for the high-scoring sessions were plotted, as illustrated in Figs. 6.6, 6.7, and 6.8. The entropy dropped in the middle and increased toward the end. Figures 6.9, 6.10 and 6.11 show the second-degree polynomial fit of Figs. 6.6, 6.7 and 6.8 respectively. The entropy in a given window will be highest when the links are most random, that is, half linked and half unlinked. Full links or empty links will result in zero entropy, so there are two reasons for an entropy drop: saturation of links or sparsity of links. Reviewing the linkographs, all could be attributed to the second reason. So for the high-scoring sessions, the beginning and the end contained more links within the 28-move window.

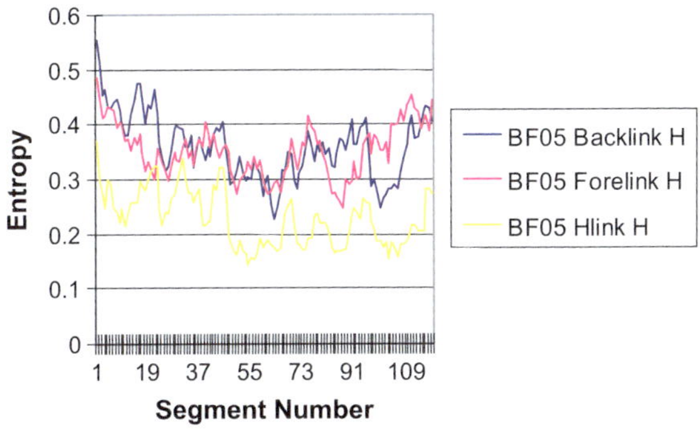

Fig. 6.6 Entropy variations in design session BF05 with total score of 7.1

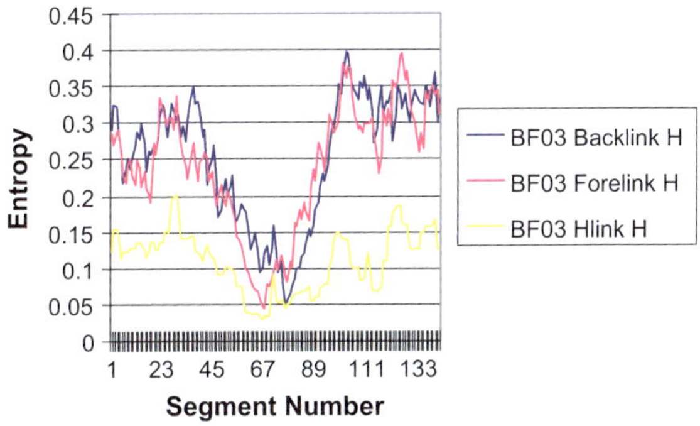

Fig. 6.7 Entropy variations in design session BF03 with total score of 6.9

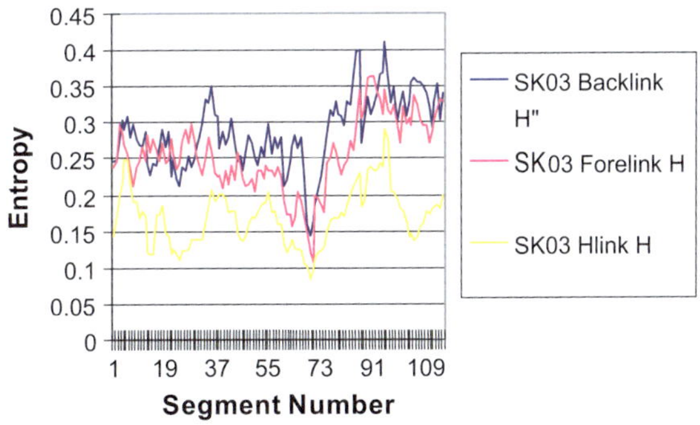

Fig. 6.8 Entropy variations in design session SK03 with total score of 6.5

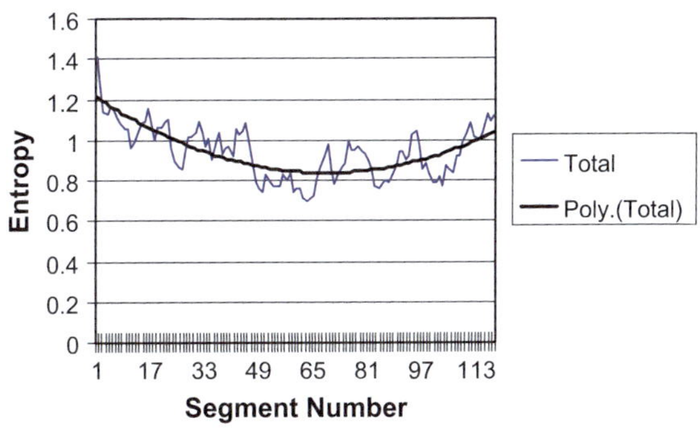

Fig. 6.9 Quadratic fit of entropy variation in design session BF05

Fig. 6.10 Quadratic fit of entropy variation in design session BF03

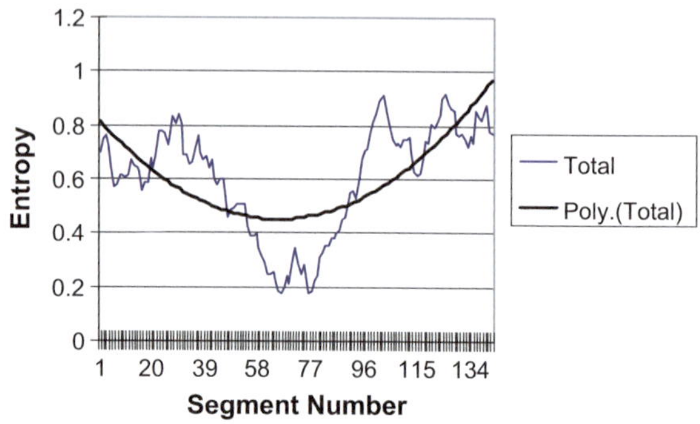

Fig. 6.11 Quadratic fit of entropy variation in design session SK03

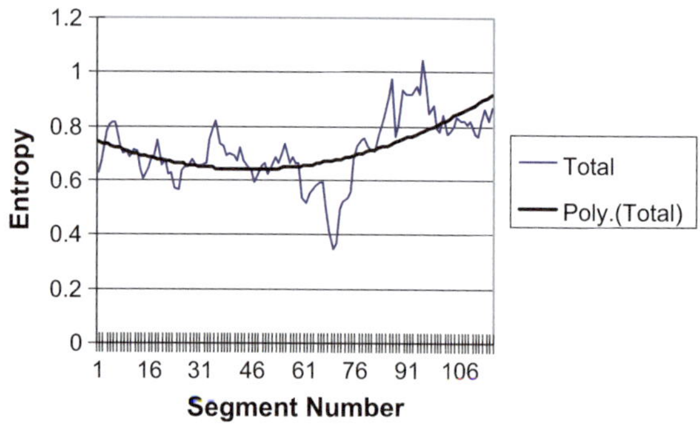

Figures 6.12, 6.13 and 6.14 are the plots of the entropy variations in the low-scoring sessions. Figures 6.15, 6.16 and 6.17 show the second-degree polynomial fit of Figs. 6.12, 6.13 and 6.14 respectively. The curvature of the polynomials were opposite to the high-scoring session, at the beginning and towards the end there were less saturation of links within the 28-move window, which meant less coherence of ideas at the beginning and end of the session.

Fig. 6.12 Entropy variation in design session SK04 with total score of 4.3

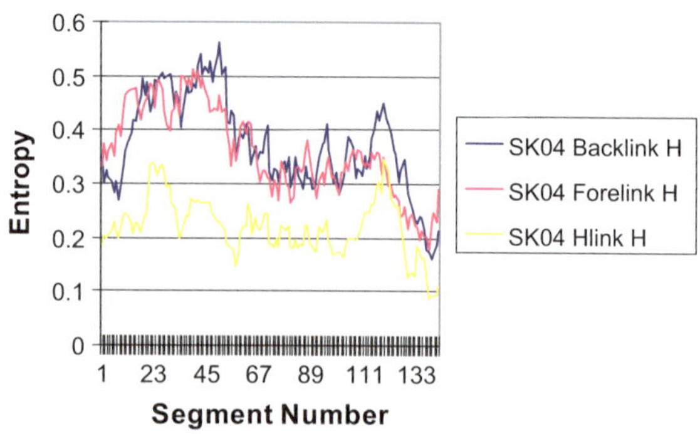

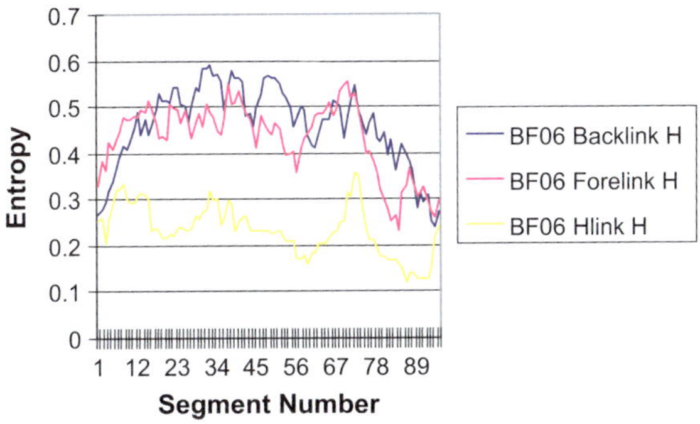

Fig. 6.13 Entropy variation in design session BF06 with total score of 4.9

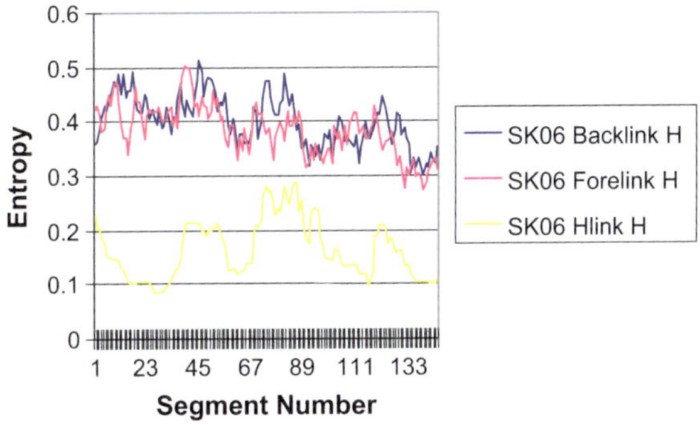

Fig. 6.14 Entropy variation in design session SK06 with total score of 4.9

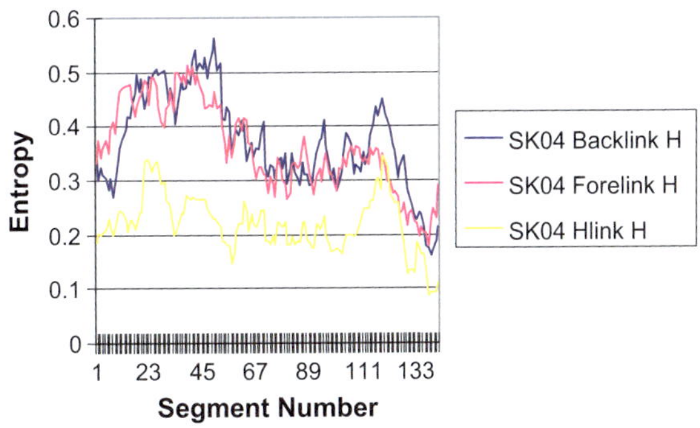

Fig. 6.15 Quadratic fit of entropy variation in design session SK04

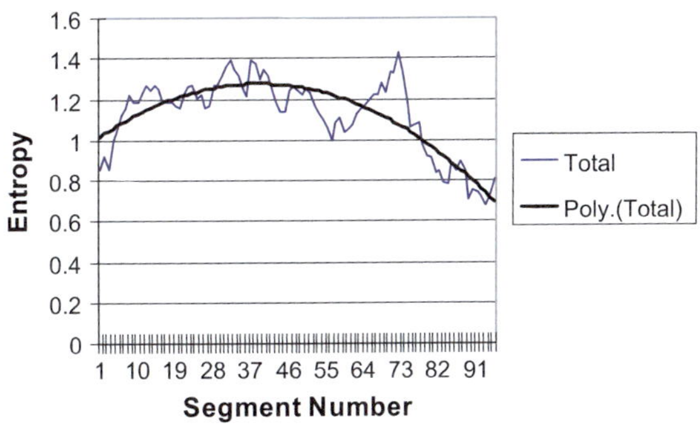

Fig. 6.16 Quadratic fit of entropy variation in design session BF06

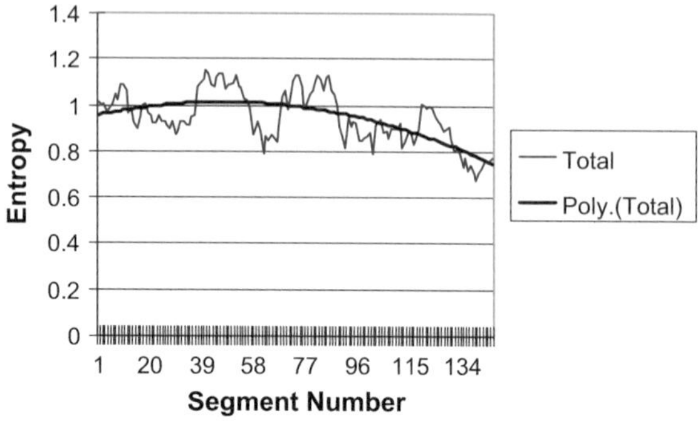

Fig. 6.17 Quadratic fit of entropy variation in design session SK06

All the high-scoring sessions have concave-shaped or negative curvature in the quadratic fit curves and all the low-scoring sessions have convex-shaped or positive curvature curves. The differential (slope or tangent) of the quadratic curve will be a straight line. The differential of the entropy curve denotes the rate of change in entropy. Figures 6.18, 6.19 and 6.20 show the differentials of the top three sessions and Figs. 6.21, 6.22 and 6.23 show the differentials of the bottom three sessions. They were plotted using the same scale.

The slopes of the differentials, i.e., rate of change in entropy, in the top three sessions are all positive. The change in entropy increases from negative to positive

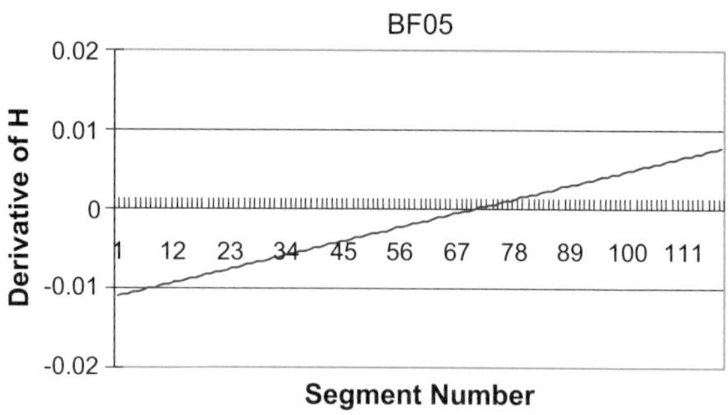

Fig. 6.18 Differential of the BF05 session

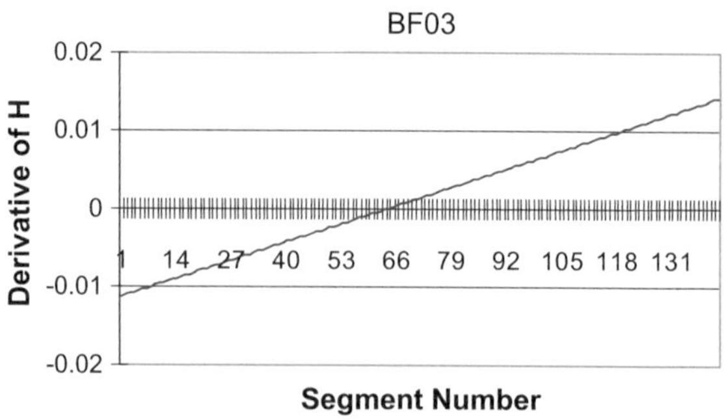

Fig. 6.19 Differential of the BF03 session

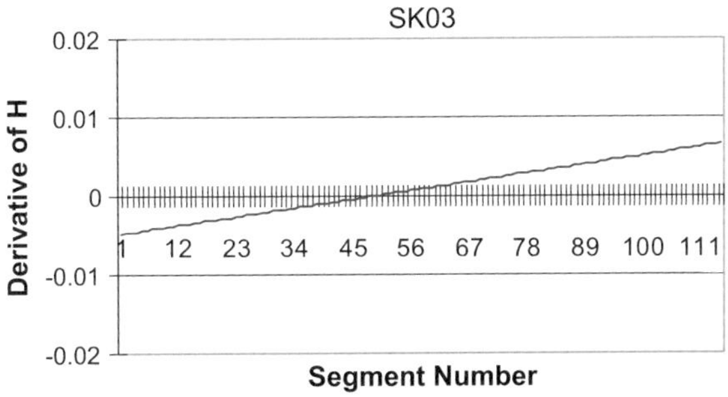

Fig. 6.20 Differential of the SK03 session

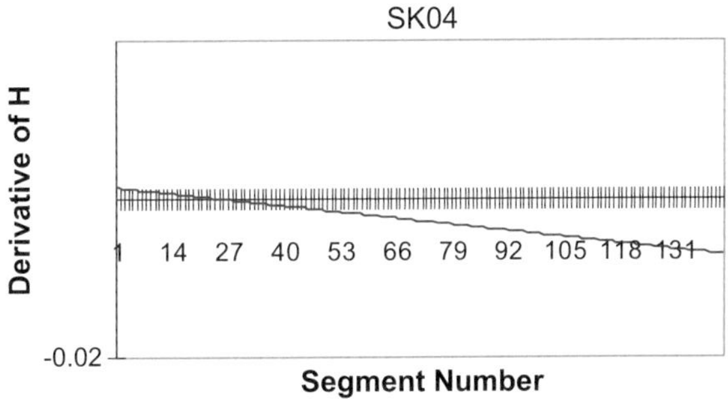

Fig. 6.21 Differential of the SK04 session

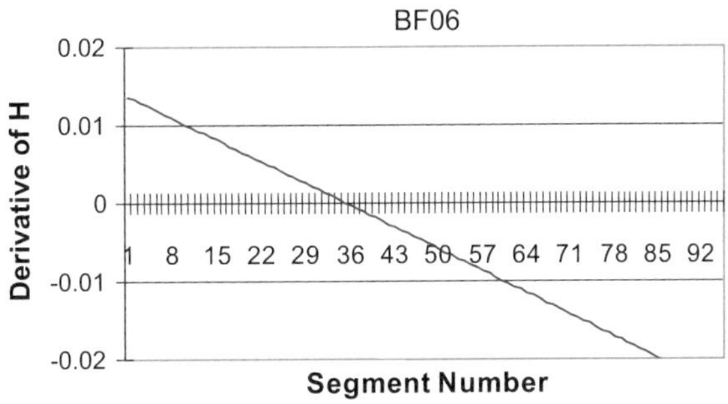

Fig. 6.22 Differential of the BF06 session

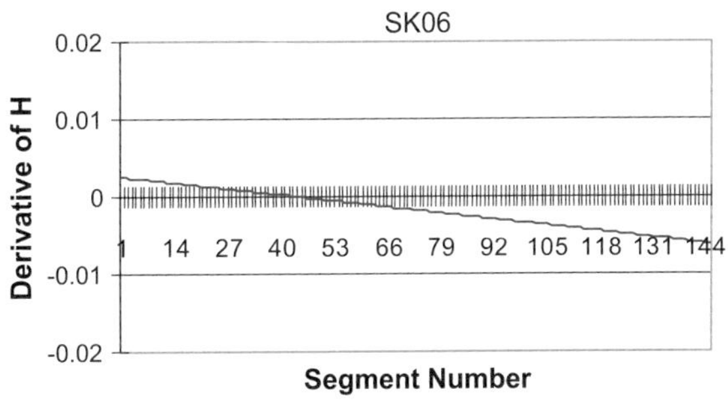

Fig. 6.23 Differential of the SK06 session

and becomes zero near the middle of the session. This is the opposite of what occurs in the low-scoring sessions, where the change in entropy decreases from positive to negative, and the slopes of the differentials are all negative.

For the high-scoring design outcome sessions the change in entropy increases from negative to positive while for the low scoring design sessions the change in entropy decreases from positive to negative within the 40 min session. A negative slop of the quadratic fit curves indicates the trend of getting less links while a positive slope suggests there is a move towards more links. In Figs. 6.18, 6.19 and 6.20, the high scoring sessions, the positive-sloped graphs cutting through zero suggest there was a change in the trend of linkages. In the first half (negative slop) of the session they started with more links and move toward less links and then changed (entropy equals zero) from fewer links to more links (positive slop). In comparison, in Figs. 6.21, 6.22 and 6.23, the low-scoring sessions, the trends were from moving towards more links and then moving towards less links.

6.7 Idea Contributions

Many researchers believe that there are prerequisites for the creation of useful ideas; among those experience and interactions play important roles. A good design idea not only fulfils the requirements but also has the quality of novelty and creativity. Finke et al. (1992) considered creativity not as a single unitary process but a product of many types of mental processes collectively setting the stage for creative insight and discovery. If good design ideas exist, bad design ideas co-exist in relative terms. Bad ideas are those that are impractical, non-innovative, or unrealistic.

A move in a linkograph does not have any attributes nor does it have any value judgment assigned to it. It is assumed there are two fundamentally different types of moves, contributing and non-contributing moves, and that contributing moves build up good design ideas and non-contributing moves do not add values to good ideas. An integration of contributing moves will produce a good design outcome and an integration of non-contributing moves will not produce a good design outcome. If a design solution is based on poor assumptions or unrealistic expectations, the outcome would not be ranked highly.

SK04, the lowest-scoring session, was considered to be the most impractical design, as seen in Table 6.1. It has a relatively high linkograph entropy, listed in Tables 6.3 and 6.4. Could this designer have integrated her/his design based on some bad ideas or non-contributing moves? Reviewing the protocol, this designer made two major decisions very early without exploring other possibilities; (1) s/he decided to have a two-storey building, (2) s/he decided to centralise the building in the middle south side of the site. From the qualitative analysis, the designer of SK04 did not have a strong central idea for the design; 28 min were spent on revisiting the structural aspect of the design. This agrees with Bilda's (2006) description of this session. He identified eight episodes in this session; four of them consisted of revising and relocating spaces. These did not seem to be contributing to good ideas besides reworking on the dimensions and locations. On the contrary, in the BF05 session, the highest-scoring one, the designer came up with many useful and practical ideas. For example, because the functions of the two studios are different, they will create different actions and sounds, which prompted the designer to separate the two studios; the analogy of the lighting behaviour of the Glasgow Art School studio space induced a window capturing the southern light; the idea of a courtyard together with the site imposed one long line of the studio spaces. Some other ideas that were not described in the qualitative analysis were: the garage roof that captures water, neighbours' and council's concerns about shadow, and a pavilion form.

If the FBS ontology were used to describe the SK04 session, one would expect that a lot of those segments would be coded as structure in the FBS ontology. Table 6.5 presents an example of those revisits of structure. The structure of the bedrooms were proposed in Segment 30 so it was coded as structure (S). In segment 66 the designer was not satisfy with the appearance (behaviour) of the facade so s/he started changing the location (Segment 67, coded as S) and proportion (Segment 68, coded as S) of the bedrooms. This kind of relocation and re-dimensioning did not introduce new variables. Out of the eight identified episodes, four of them contained this kind of structure revisiting.

Those ideas in the BF05 session would expect to be coded ranging from function and behaviour to structure. One would expect the types of reformulation to be richer in the BF05 session. An excerpt from BF05 with the FBS codes is presented in Table 6.6. The link from Segment 53 to Segment 52 is a type II reformulation.

Table 6.5 Examples of FBS coding for SK04

Segment	Transcript	Links	FBS code
30	The master bedroom is 20–25. I've forgotten the master bedroom. Okay, so that's another… I'll add that to my list. 20–25 m^2. Okay	29 28	S
66	Now I'm not quite happy with the amount of area I've used because it presents a very solid facade. I'd prefer to reduce the amount of bulk on the top floor at the front and shift some of this back	65 30 11	B
67	I could possibly also move some bedrooms around so that I have more of these facing the north and looking over the open space and not having two looking over the southern boundary. So I can see from what I have now if I take those two rooms, they could	66 30	S
68	If I reshape those proportions slightly, I'll draw them in approximately—6 m deep. Then I'll cross out the ones at the front and move them to the rear. That's bedrooms 3 and 4, I'll keep the same numbers	64 65 66 67	S

Table 6.6 Examples of FBS coding for BF05

Segment	Transcript	Links	FBS code
49	And that would also work for the art studio because it is on the facing east. Facing to the park. So it gets the morning sunlight, because you don't want to be in the cold	12 21 11 48	B
50	dance studio and then just getting westerly sun without the need really to be controlled so it don't get too hot	21	B
51	Ok, so that could work, your loo could go there so it would be a linear thing up there, feed off to, that could work. I am not quite sure about the size of the spaces but	43 46 42	S
52	I think it would be like an H plan and then you come into the centre, you would spread either way. Whether the circulation is becoming too much on the house. Maybe you don't do that. Maybe you have…	32 31 25	S
53	maybe you don't have a roof… or you just have an open connection or walkway. That could be quite nice. So your living spaces then connect onto that. So from your kitchen you could see right through	17 37 52 47 48	B
54	And then do you… you make the site, you don't have a specific garden, you have a big park. I get a sense that she is a painter, they are not real gardeners, but they may have…	25 31	F

From the analysis of the "H" shaped plan and the amount of circulation space prompt the designer to have the expected behaviour of open connection. This open connection was a new behaviour variable.

6.8 Conclusions

In the design sessions presented in this chapter, the design outcomes had certain relationships with the entropy of their linkograph. All the blindfolded (BF) sessions had a higher or equal score for the design outcome compared to their corresponding sketch (SK) sessions. This was reflected by the entropy of the linkographs: five out of six of the BF sessions had higher entropy than the corresponding SK sessions. However, higher entropy did not correlate with better design outcomes. All the high scoring-sessions had higher entropic measures toward the end of the design sessions as they become more integrated approaching the end. This was approximated by a quadratic fit with a negative curvature of entropy curves. The differentials of these quadratic curves yielded straight lines. The three highest-scoring sessions had a positive slope, while all the poor-scoring sessions had a negative slope. The decrease in entropy at the beginning of the good sessions was caused by fewer connections between moves, which could indicate diversification of ideas. The increase in entropy at the end of a session meant a better integration of moves, which might indicate a consolidation of ideas. Does this mean that good designs need to diversify before consolidation? Is the change in entropy an indicator or a predictive tool for good design? With these limited cases, a firm conclusion cannot be reached. However, there are some indications in this case study that there is a correlation between the inter-segments entropy variation and the design processes and/or its outcome; the trend of entropy variation may reveal the outcome of the design. More experiments are needed to verify this claim. There was some evidence that suggested the design outcome could be related to the entropy of the linkograph. Also, FBS ontological coding is expected to help in identifying good design sessions, and this will be discussed in the next chapter.

Chapter 7
Conclusions

This concluding chapter discusses the usefulness and shortcoming of the techniques described in this book.

7.1 Design Protocol Studies

The aim of the research reported in this book has been to develop and test new approaches to the quantitative analysis of design protocols that will increase our understanding of designing.

In research studies undertaken in many established fields, such as science and medicine, the modes of inquiry are well developed, with clear methodology. However, in the methods used to study designing, there is a lack of uniformity. This study has striven to find an approach that is general enough to cover most design scenarios but specific enough for the study of designing.

"Protocol analysis lies in the middle ground between the experimental methods of the natural sciences and the purely observational methods of the social sciences" (Cross 2007b, p.ix). Most protocol studies of designers do not have an adequate amount of data to reach statistically significant results due to the limited number of participants.

Some of the results presented in this book may fail to draw solid conclusions. Nevertheless, the relevance of this work lies in providing alternative methods of observing design protocol, hence bringing new insights into design cognition. It uses developed mathematical methods to analyse protocol data. These methods include Shannon's entropic measurement of the linkographs, combining ontologically based coding with linkography, and Markov analysis of an ontologically based coding.

J.W.T. Kan and J.S. Gero, *Quantitative Methods for Studying Design Protocols*, DOI 10.1007/978-94-024-0984-0_7

7.2 Reflections on Linkography

The advantage of using linkography are fourfold. Firstly, it captures aspects of both process-oriented and content-oriented aspects of designing. Secondly, it is scalable in two dimensions: this method is not tied to the number of designers being studied (Goldschmidt 1995) and the length of the linkograph can be of any duration. Thirdly, it is flexible, the design moves and how the design moves are linked can be coded separately, depending on the focus of the study (Dorst 2004; Kan and Gero 2004; van der Lugt 2003). Fourthly, it can be used at different levels of granularity, at the cognitive level or at a coarser level of ideas or decisions. Using Goldschmidt's example in Chap. 3, Fig. 3.2 was segmented with a coarser grain while Table 3.9 shows the same protocol with a finer grain. Comparatively, the segments in Chap. 6 has a finer grain than the segments in Chap. 5.

Linkography does not capture all the aspects of design activities. As with any protocol study, it relies on inter-coder arbitration (McNeill et al. 1998) to ensure protocols are objectively segmented and coded. In constructing a linkograph van der Lugt (2000) has attempted to strive for reliability in linking by introducing a series of indicators for links. The use of ontological coding has a similar effect because when deriving the processes the analyst has to reason why the segments should be linked.

All the cases in this book were studied using FBS coding or linkographs or both. The focus was on acquiring quantitative information from those graphs. The statistical analysis of the linkographs produced quantitative results that was not available previously. In Chap. 4, commercial software was able to generate clusters automatically. These clusters correspond to the actual design activities, hence the semantics of a cluster could be labelled. Entropic measurement provides another way of benchmarking design processes from linkographs. After classifying the segments, the links of the linkographs can be viewed as transformation processes, which quantifies the design session in terms of ontological processes.

7.3 Ontological Coding of Linkographs

The processes from the FBS ontology are claimed to be fundamental for all designing. Unlike most coding schemes which allow overlapping of codes, the ontological approach requires discernment of one code per segment. This clear distinction converts the protocol into unambiguous segments; it quantifies the amount of cognitive design effort spent in relation to function, behaviour, or structure. The links not only provide a structural view of the processes but also provide an opportunity to locate the frequency of each design transformation process. This nested representation of FBS-coded linkographs provides an opportunity to examine the design protocol not in a linear manner but as a network of processes.

The results presented in Chap. 5 have demonstrated that this approach can be used to study teams and multi-disciplinary teams of designers, two architects designing face-to-face, and one architect and one landscaper designing in an Internet-based 3D environment. Other studies (Goldschmidt 1995; van der Lugt 2003) have shown that linkographic studies can be applied to one or more designers. These provide a common platform to analyse and compare design activities in various circumstances.

In a group setting, the situated FBS coding mainly captures design processes; other processes such as social and management processes are not directly captured. However, when examining the linkograph, individuals' contributions and the interaction among the members can be observed. This might reflect some of the social processes. For example, if an individual does not get responses from others, his/her link density will drop.

Coding with situated FBS is able captures cognitive actions and processes that FBS cannot capture, for example the re-interpreting of state variables and focusing on aspects of a state variable. In order to successfully apply the situated FBS coding scheme, the granularity of the segments needs to be very fine. Reviewing the segments and the codes of the face-to-face design session in Chap. 6, the granularity was observed to be too coarse for this purpose. For example, the drawing actions had been segmented separately but not the gesturing of space and objects. Consider this incident in the protocol concerning the discussion of the entrance: "Through there or around the icon? (point and gesture around the circle)" had been treated as one segment. In this segment, a new concept—the notion of an "icon"—was introduced. However, it was coded as expected structure because of the suggestion of location. Retrospectively, it should have been separated into two segments. The "icon" should have been coded as "function" and the gesturing as "expected structure".

The FBS and situated FBS ontologies provide ways to investigate design activities. An examination of the examples indicate that segmentation and coding require training. Typically, coding tutorials involve working with an existing transcribed, segmented and coded protocol that has already been arbitrated. Coding tutorials commence with the learner coder being introduced the FBS ontology. Then they are given the first few minutes of the transcription of the video of the protocol and asked to segment them. They then compare their segmentation with the arbitrated segmentation up to that point and discuss with the tutor the differences. If there is no tutor they have to attempt to understand the differences. They then code these segments and compare their coding with the arbitrated coding and attempt to understand the differences. They then segment and code the next 10 min of the transcript and compare their results with the arbitrated results. This continues in 10-min chunks until the end of the protocol. The comparison of the per cent agreement between the learner coder's results and the arbitrated results quickly improves to reach over 70 % agreement. Two coders working independently and then arbitrating between themselves produce agreements of around 75 % after a few protocols have been coded.

On applying the proposed ontological coding scheme, one of the weaknesses in this study was not to distinguish when new FBS variables were introduced from when only the values of variables were changed. New variables are introduced when there is a reformulation process. The links between revisited/repeat moves were not discerned from the real reformulations. For example, many revisits of structure, as seen in the lowest-scoring design session in Chap. 6 did not introduce a new structure variable but rather changed the value of the variable. There are lessons to learn from van der Lugt's (2003) approach of link types. Tangential link type suggests a different direction of ideas is essentially introducing new variables, which should be considered as a reformulation process. Supplementary link type make small alterations to the original idea should not be considered as reformulation. Modification link type change the original idea in the same direction should be considered as a focusing process. However, once a protocol has been coded using the FBS ontology it takes relatively little effort to examine each segment and add a supplementary code that distinguishes new variables from existing variables without changing the existing coding (Gero and Kan 2016). Figure 7.1 shows the results of coding "new" and "surprising" segments in a protocol that has already been segmented and coded using the FBS ontology. From results such as these it becomes possible to use the FBS-coded segments as a framework for other kinds of supplementary coding (Yu et al. 2015; Song et al. 2016).

This methods in this book have not fully capitalised on the situated FBS framework. The push-pull and the focusing processes have not been presented (Gero and Kannengeisser 2004). The assumption in deriving the processes from linkographs was that it is a forward transformation process (forelink) derived from looking back, as illustrated in Fig. 7.2.

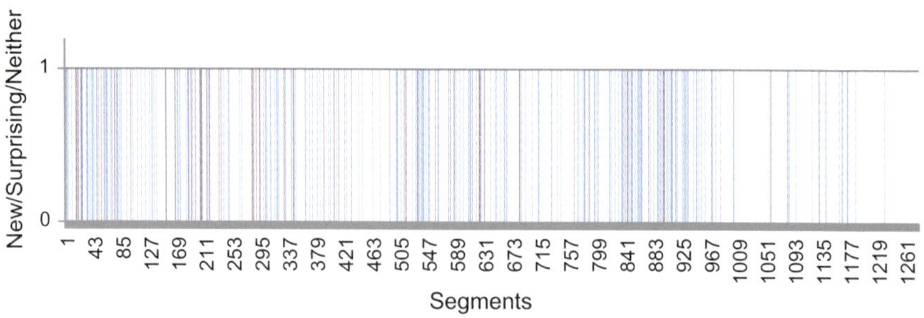

Fig. 7.1 New (*blue*) and surprising (*red*) segments in a protocol, white means neither new nor surprising

Fig. 7.2 Deriving processes from links as a forward process

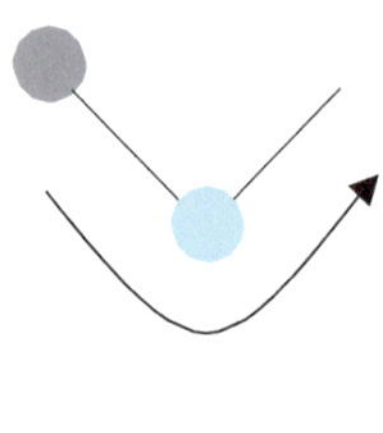

The push-pull process, based on constructive memory, is not only a forward transformation process but it is also backward transformation, modifying memory (Gero 1999; Riegler 2005; Schacter and Addis 2007). The meaning of a depiction changes along with designing. Using the example of the "icon" again, the drawn circle's meaning changed from the possible location for an entrance to an icon to capture attention. In order to fully capture this the situated FBS framework is needed to represent this cognitive process. Recent research is starting to bring design cognition closer to the foundations of cognition (Hay et al. 2016), which points to a direction for future analysis of protocol data.

7.4 Markov Analysis

Markov analysis, taking time into account, provides a way to examine design protocol data in a sequential manner. It does not assume successive events to be independent. Gottman and Roy (1983) claimed that "Anyone who has collected data over time and ignores time is missing an opportunity" and "The dimension of time is so central to conceptualizing social interaction that its use will lead us to think of interaction itself as temporal form".

In this book Markov analysis was the only analysis that was independent of the linkographs. A first-order Markov chain was used to model the design process in terms of the sequence of the situated FBS events. The transition matrix can be viewed as a signature that summarises the transitions between all the FBS events.

The mean first passage time matrix provides yet another view of the design process. The calculation of first passage time can be used as a tool to test hypotheses such as: "it takes longer to get from F to S than from B to S". This hypothesis is supported in all the cases studied here.

From the results of Chaps. 3, 4 and 6, the values of the probability vectors are very similar to the statistical value; especially when the number of events is large.

The dimension of a situated FBS-coded protocol Markov chain will be a 10 by 10 matrix, which is difficult to comprehend and manipulate. Some of the equations in Chap. 6 were solved with the assistance of MatLab. The amount of data required for a higher order of Markov chain analysis with 10 categories is large.

Similar to the entropic measurement, the Markov analysis of design protocol is also in its infancy. It can be applied to any coding scheme. Although its capability and applicability still need further exploration, the results presented here are promising. The first passage time can be used as a hypothesis testing tool. The next step is to determine the order of the Markov chain instead of assuming the first-order chain. Gottman and Roy (1983) used information theory together with the likelihood ratio Chi-square for testing the order of a Markov process. The goal is to understand the design process in terms of the sequential pattern of FBS states. After that, a stationary analysis should be performed. This involves dividing the protocol data into time periods and compares the transition matrices. If there is a stationary chain, the stationary distribution can be regard as the DNA that governs that design process in that design session.

7.5 Entropic Measurement of Linkographs

The theoretical background for entropic measurement of linkographs was covered in Chap. 3. In the pilot study (Chap. 4), the two design sessions were of very different natures; the entropic measurement matched the qualitative analysis. Forelink entropy reflected the idea initiation opportunity and backlink entropy indicated the opportunity of building upon old ideas. The measurement of horizonlink entropy was inconclusive for these design sessions. The individual entropy score matched the observed role and participation of the individual designers. The total cumulative entropies agreed with the link index with a different magnitude.

The experimental results presented in Chap. 5 hinted that the design outcomes had certain relationships with the entropy of their linkograph, especially the change in entropy within a design session. One would expect that the design outcome would positively correlate with the entropic measurement of the linkograph. However, statistically, the design outcome did not correlate with the cumulative entropy. Looking at the lowest-scoring session with high entropy suggested that linking segments with slight modifications does not contribute much to the quality of the design. More systematic investigations are needed to examine if the types of links, as reported by van der Lugt (2003), will affect the design outcomes.

The quadratic fit of entropy variation in the high-scoring design sessions have concave-shaped or negative curvature in the curves and all the low-scoring sessions have convex-shaped or positive curvature curves. This may be used as a predictive tool for good design outcomes. This needs further investigations because of the limited number of cases.

Krippendorff (1986) applied information theory, entropic measurement, to structural modelling for qualitative data and compared it with network analysis, path analysis, Chi-square, and analysis of variance. The concepts of structural modelling depicted by Krippendorff (1986) can be applied to design protocol data, which may provide alternative ways to analyse design protocol.

The cases presented in Chaps. 3, 5 and 6 have demonstrated the potential of using entropy as a tool to investigate design protocols.

7.6 General Discussion

The development of a scientific understanding of design requires empirical data from designers designing on which to found and test models of designing. A scientific understanding is based on the axiom that the phenomena being studied can be observed, ie, sensed in some way and that there is a regularity associated with it. A scientific understanding of design does not imply that design is necessarily a science only that it can be studied in a certain way. One method of collection of that data is through the study of design cognition. Most models of design assume that designing is a process, rather than being some mysterious activity, and as a consequence design cognition can be studied scientifically. Progress in design cognition research has been hampered by the inability of researchers to build on the work of other researchers. This has been caused by a lack of commonly used methods and commonly agreed analytical tools producing a lack of commensurability of the results. Whilst there are many ways of viewing designing the claim is made that the fundamental issues and processes involved in designing are not uniquely related to any particular design task, designer or design situation and the issues and processes can be studied independently of the design being produced (Asimov 1962; Coyne et al. 1990; Dieter and Schmidt 2008; Dixon 1996; Dym 1994; Eggert 2002; Eide et al. 2001; Ertas et al. 2008; Gero 1990; Gero 1991; Gero 2008; Hatamura 2006; Lawson 2005; Matthews 1998; Rychener 1988; Ullman 1992). This is not to imply that there is only one way to carry out research into design cognition, rather it is suggested that within a research paradigm if there is no commonly used approach to utilizing and measuring the source data it is difficult for one researcher to build on the results of another researcher and that this impedes progress.

There are measurement concepts that have not be included in this book that are currently being explored. These include for example treating a linkograph as a network graph and studying the centrality and betweenness of issues, and methods of translating between coding schemes.

This book has aimed to bring to attention of design researchers two strands of design cognition research: methods of segmentation and coding and quantitative methods for studying segmented and coded protocols. It is hoped that this small volume will arouse interest in unifying design cognition research.

Appendix A
Samples of Coding

A.1 Chapter 4: NetMeeting

In column 1 (Subject), A represents the architect and E represents the landscaper

Subject	Transcript	Segment	Code	Links
A, E	(looking at brief)	1		
A	I've no idea how we can actually get this to work in terms of the area, we have nothing to scale with	2	b	1
E	How do you scale, yea	3	s	2 1
A	how the hack can we do	4		3 2 1
E	we do approximate of the … this is scale on the this paper	5	s	4 3 2 1
A	sure, but there is nothing on this things to tell us what that would be, it'll be awfully rough	6	s	5 4 3 2 1
E	it says 68 m, just have to rough	7	s	6 5 1
A	Now what … normally if we you would have had Arhicad open or something to allow you to do this	8		7 6 4 2
A	it is possible, [z Yea but it crashes with netmeeting] [E it crashes] does it	9		8
A	So what we are doing here is just theoretical, I don't think we can actually achieve those areas without even thinking about them more seriously so … the exercise is communicate and try to do a design session … assuming what we are drawing will become the area we are saying,	10	f	8 7 6 5 4 3 2 1
A	fine … it is just like a mock up, cause it really doesn't ha have a way of doing it do we [z yea]	11	f	10
E	so we just get the em … the appropriate function relationships	12	f	10 6 5 4 3 2 1
A	yeh, that's what I think too, yea	13		12

(continued)

J.W.T. Kan and J.S. Gero, *Quantitative Methods for Studying Design Protocols*, DOI 10.1007/978-94-024-0984-0

(continued)

Subject	Transcript	Segment	Code	Links
E	approximate sizes	14	s	13 12
A	yea	15		14 12
A	(looking at brief)	16		
A	so we're talking about an art gallery (looking at the brief)[yeah (looking at the brief)]	17	f	16
A	(looking at the brief) and the main the permanent exhibition is 1500 so that is the biggest piece,	18	b	16
A	a scripture garden 600 and we've 2 temporary areas one large one small	19	f	16
A	and a forum. I don't know what forum means, (flip to next page of brief) it's a central lobby maybe? I guess	20	f	16
E	… like a leisure space (flip to next page of brief) or a discussion space	21	b	20
A	it could be could be a theater (flip back to first page), yeah.	22	b	21 20
A	And then you've got the entrance foyer which hasn't got anything	23	f	16
A	reception 30 m, cloak 20 (reading from brief) cafe 200, shop 100 … storage space for the shop (looking at brief) etc … so this is base on something … I guess [… we got]	24	f	16
A	the north is up the page (turn to start looking at display) so the best sun is down the page	25	b	16
E	(looking at brief)	26		
E	for the gallery spaces	27	s	26 25
A	yeah, yeah(E start to pick up pen)	28		27
A	so can you see my mouse, my pointer	29		
E	can I see your mouse?	30		29
A	my pointer	31		30 29
E	(looking at monitor) yeah well I can see a pointer … is that mine?	32		31 30 29
A	that's your but you don't see me moving it	33		32
E	no	34		33
A	ok that mean I've to draw	35		34 29
A	alright, so What I would say we could do is to think of first of all how we can have access to this ah if we're to think of the area (looking at display, holding mouse and drawing line with mouse) [per meter????]	36	f	
A	I see the major area for from here It think the main area for … (draw)	37		36 27 25
A	Can you see that now?	38	s	29

(continued)

(continued)

Subject	Transcript	Segment	Code	Links
E	yeah can see your drawing yeah	39		38
A	so this maybe the area given to the main exhibition area because	40	s	37 18
A	even though we've the foreshore in that location of the north	41	b	40 37 36
A	that would perhaps be the presentation area to the to sea	42	b	41 40
A	its a main square like in this area here which maybe the preamble,	43	b	42 41 40
A	we can celebrate the enter to the gallery by mimicking a similar space across that	44	f	43 41 40 37 36 27 25
E	(looking at brief) is there a square of building	45		43
A	it looks like a square see how it has its shadow [yeah] because the North east is up the page I think that is the shadow	46	b	45
E	so … shadow of this building, so this is an open space	47	b	46 45

A.2 Chapter 6: Brainstorming Session

In column 1 (Subject) number 1 through 7 are used to represent the seven participants. The corresponding coding are re:R, fi:F^i, fe:Fe^i, xf:F^e (not found in this example), bi:B^i, be:Be^i, xb:B^e, si:S^i, se:Se^i, and xs:S^e.

Subject	Time	Transcript	1st	2nd	Arb	Segment	Links
1	0:05	quite important is its about the thermal-incli- inclis () pen	re	re	re	1	
1		but alo about the media	re	re	re	2	
1		so it's the **interaction** between the special paper and the thermal pen basically	be	bi	bi	3	2 1
1		**focuses** about the print	fi	fi	fi	4	1
1		the paper is involved	si	si	si	5	2
4	0:06	the one on the top left is the one that's the real main focus of the of what we're doing today [AJ:yeah]	re	re	re	6	5 4

(continued)

(continued)

Subject	Time	Transcript	1st	2nd	Arb	Segment	Links
		that's the standard plain thermal paper err					
4		and then **it can draw**	fi	fe	fe	7	5 4
4		either atoms or line types	be	be	be	8	7
4		so for the process of making this prototype that's going to help us sell the project that's the concept really err	o	o	o	9	1 7 8
4		design a-a prototype	re	re	re	10	9
4		the other ideas on the paper there were other concepts for a thermal pen which erm may form the ultimate product	re	re	re	11	3 2 1
4		the main focus is that style pen where you'll **create a set of patterns** by moving on plain white paper	be	be	be	12	3 7 8 5 4 3
5		will this be in colour?	bi	bi	bi	13	12 11 8 6 1
2		yeah well we should …	be	bi	bi	14	13
2		we **can print thermo reactive dyes** onto media substrates	bi	be	be	15	14 13
2		and then when you heat them up they **cool to reach the colour** they're supposed to be so erm we've done some test prints on this erm been tested () by the HP () boss	be	bi	bi	16	15 14 3 2 1
2		playing with erm hot air () miraculously a colour image appears erm out of a blank sheet of paper	bi	be	be	17	16
2		74 so providing we could **provide a heat source**	se	se	se	18	17
2		then we could reveal anything which is printed on the paper	bi	bi	bi	19	18
2		76 technically that's more demanding because energising the entire () printout takes time hammers the batteries and its quite a demanding thing to do	bi	be	bi	20	19
2		78 so technically the stuff where you've got a **low percentage fill**	be	bi	be	21	20 19 17 16 15

(continued)

(continued)

Subject	Time	Transcript	1st	2nd	Arb	Segment	Links
2	0:08	sort of faces and text and things like that is a lot more **realisable**	be	be	be	22	21 20 12
2		but if we could do that with the media whether we could do that with this adaptation I don't know	si	se	se	23	22 21
4		the one on the right- the one on the right **doesn't actually have to be a thermal pen** () it could just be-	bi	be	be	24	16 17 19 20 21
1		it could be **a soldering iron**	si	se	se	25	24 18
5		or a hot air ()	si	se	se	26	25 24 18
1		but basically th-the focus is on that at the moment and basically th- the idea is that there is a thermal print out in **in the nib** of the pen	si	si	si	27	26 1 25
1		whichever shape that takes now with that concept ….	si	si	si	28	27
		there are a number of specific problems that need to be overcome and I think that's what we want to focus on today in the mechanical brainstorm and then on Monday there's an electronics brainstorm as well errm that will cover a range of other topics so	o	o	o	29	
1		basically the first problem that we need to think about is the **wobbly** arm movement of the users …	re	re	re	30	
		this is aimed at potentially but not exclusively at five to eleven years olds who have some writing skill but not fantastic writing skill	o	re	re	31	30
1	0:09	they're likely to hold the pen different ways erm different angles	re	bi	bi	32	31 28 27 30
1		etcetera and so ther-there's two important things one is…this concept one is that that the printout needs to be this **contact maintained**	re	be	be	33	32 31 30 28 27
1		the printout needs to stay in the **right angle to prin**t…even though the user is not capable of keeping exactly the right shape or weight so one of the things	re	be	be	34	33 32 28 27

(continued)

(continued)

Subject	Time	Transcript	1st	2nd	Arb	Segment	Links
1		…other products or situations where a product needs to follow a contour or needs to be able to adjust itself keeping itself in the same alignment even thought the user might be moving things about …	bi	be	be	35	34 33 32
7	0:10	is it only paper you're thinking about or could it be **other things** like mugs or fabric or pottery	si	bi	bi	36	2 3 5 6 15 35 34 33
4		it could be anything	si	bi	bi	37	36
4		I ended up with the + hold on + sledge	si	si	si	38	35
1		the sledge excellent so what did that generate then? (write: sledge)	xs	xs	xs	39	38
4		the sledge manages to keep level by having quite a wide base	bi	si	bi	40	38
1		(write: wide base)	xs	xs	xs	41	40
4		a main force in the middle	bi	si	bi	42	40 38
4		unlike the set of skis	bi	si	si	43	40 38
4		where quite narrow and	si	si	si	44	43 40 38
4		you go up on an edge when you're turning	bi	si	bi	45	44 43 40 38 42
4		the sledge is er quite broad	si	si	si	46	43 40 38 44
4		and then you have the weight right in the middle	si	bi	bi	47	46 43 40 38 42 40
4		so they manage to keep both runners on the snow	bi	bi	bi	48	46 45 42 38 40 47
1		(write: force in middle)	xb	xb	xb	49	47 42
4		a sledge or a snowboar- a skis or snowboard	si	be	be	50	38 47 46
1		some **guiders** almost **down the side** of this	si	si	si	51	38 50
4		the easiest way to **keep the pen at a right angle** would be	be	be	be	52	34 32
		to **have a set of stabilisers** on it based on the idea of a sledge	se	se	se	53	38 52 40 46
1		stabilisers +++like a bicycle yeah that's a good	si	si	si	54	53

(continued)

(continued)

Subject	Time	Transcript	1st	2nd	Arb	Segment	Links
1		(write: stabiliser)	xs	xs	xs	55	53
5		a flat base with a sort of **universal joint like a windsurf mast**	se	se	se	56	40 35
1		(write: Universal base….)	xs	xs	xs	57	56
5		it **stays flat** but the bit you **hold onto can be at different angles**	bi	se	se	58	56 32 34
5		the **size** of the thing in contact with the paper you could either have a **very small** contact area and er track where it goes **or** you can have a **wide contact** /area\-	si	se	se	59	58 27 33
4		show the **relative size of the pen** if you've got an example	si	si	si	60	59
2		**size is dictated by cost** really it needs to be quite narrow	si	bi	bi	61	60
1		lets not be too…lets just use that as guidance but let's- let's not be too **preoccupied** with the **shape** that is at the moment	si	si	si	62	61 60
1		we'll just look at **the tip** of that… rather than the rest I would suggest	si	si	si	63	62 61 60 27
4		it can be te- **ten mil** across	si	si	si	64	63 60
3	0:13	it's a erm its **not an edge** one then it's a it's a on the **flat** there are two flats printing aren't there more than one	se	se	se	65	63
3		(draw: ref:E1_15 (Flipchart).jpg top 2)	xs	xs	xs	66	65
3		(while drawing) if I where yeah the **resistors sit on the lump there** and you've got the type where the resistors sit on this **corner** here	se	si	si	67	66 65
4		it's a corner one	si	si	si	68	67 66
2		no its not cheap ()	bi	bi	bi	69	68 67
3		it'll be about fifty percent more expensive	bi	bi	bi	70	69 68 67
3		there's an end that goes over the paper isn't there sort of thing?	si	se	se	71	
2		(draw: ref:E1_15 (Flipchart).jpg middle overlap left)	xs	xs	xs	72	71 66
2		there's a type that does this as well though	si	si	si	73	72 67 66 65

Appendix B
Briefs of the Experiment

B.1 Brief for Chap. 4 Brainstorming Session

Project Penny: Print Head Mounting—Mechanical Design Brainstorm 10/Nov/06 CR2

Briefing for Attendees

- Introduction

Project Penny is an internal TTP project working in collaboration with Refined in China to produce a working demonstrator of a thermal printing pen.

The pen will work on thermal media and may operate as a normal pen but also contai n pre-programmed patterns or different print widths as shown. This may be a toy or a serious art product—tbc.

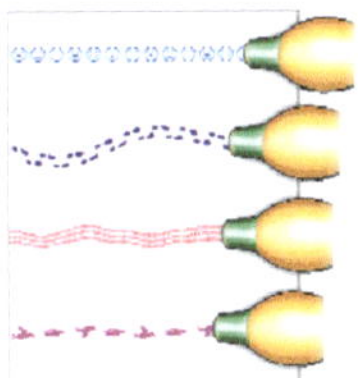

- About the brainstorm

Completely unrelated to this project, PE is participating in a study by the Open University on "Design meetings in practice". As part of this study this (and another electronics) brainstorm will be filmed and analysed. To help the brainstorm run smoothly AJ will be moderating the sessions.

Several other companies from a wide range of sectors will also have sessions filmed and the output used to produce initially a paper for a conference and ultimately a book. The film of our brainstorm will not be shown publicly and the brainstorm content is confidential—you will not be mentioned in this forthcoming bestseller. Pseudonyms will be used in any transcripts.

J.W.T. Kan and J.S. Gero, *Quantitative Methods for Studying Design Protocols*, DOI 10.1007/978-94-024-0984-0

- Brainstorm Topics

To help you look clever in front of the cameras here is a pre-brainstorm briefing of what problems we have to solve with the print head mounting design and pen format:

1. Keeping the print head level—as the user moves the pen across the page the 5–10 mm long print head has to remain in contact with the paper—what mechanism could take account of the user's wobbly arm movement?
2. Protecting the print head—if the pen is pressed (or bashed) too hard onto the surface it can be damaged—what me chanism could protect the print head from high contact forces?
3. Print head activation—to protect the print head from overheating can we design a mechanism which activates th e pen on contact with the media?
4. Print head angle—the print head will only work when presented to the media at a limited range of angles—can the pen be designed ergonomically to force/train the user only to use with these angle limits?

- Pre-Brainstorm Homework: To stimulate thinking about smooth running print head, your pre-brainstorm task is to bring along a product (or a picture of a product) that has to glide smoothly over contours.

TJM/1458

B.2 Brief for Chap. 4 Face-to-face Session

CRC Team Collaboration in High Bandwidth Virtual Environments
Face to face Page 1

Sydney University Students' Union Gallery

- THE PROGRAM

You are asked to design a contemporary art gallery to house the Sir Hermann Black Collection and to provide space for temporary exhibitions for the University of Sydney Students' Union, which is being expanded, and is being moved from the the Wentworth Union as a result of the University of Sydney's Masterplan.

- YOUR CLIENT

Your client will be the University of Sydney Students' Union, represented by Mr Nick Vickers, who is the Director of the new Union Gallery and the Director and curator of the Sir Hermann Black Gallery currently housed on the top level of the Wentworth Union.

- THE SITE

The site is on the corner of City Road and Darlington Street, and is part of the Darlington campus. The site is currently occupied by the University Regiment and is defined on the accompanying plans.

Approval for any building on this site would be influenced by the University of Sydney Masterplan, and governed in the regulatory environment by the Local Environmental Plan and the Development Control Plan of South Sydney Council.

We have provided you with 2 × A3 of the site at 1:500, and 1:200

Approval and regulatory requirements

The site area is approximately 1800 m^2.

The maximum site coverage is 33 %. That is, 594 m^2, the building 'footprint', or built coverage of the site, excluding any external courts, sculpture display areas, unloading docks, service areas, etc. Building height limit is *three storeys*, and the building height and mass should enhance the urban context and respond appropriately to surrounding buildings and streetscape.

- THE BRIEF

Discussions with Nick Vickers, the Director of the Union Gallery and advice from Jan Feildsend, architects Paul Berkemeir and Colin Still have produced the following general brief.

The Union Gallery—the client's aims

The Union Gallery will encourage the public to enjoy and engage with the art works in the exhibitions of both the permanent collection and the temporary exhibitions, and will present the works in the best way possible.

The Union Gallery is to have a *community focus*. The major aspect of this will be the connection made by the Gallery and the art it houses with the University and wider community, both in terms of the physical fabric of each, and in terms of the varying communities of people in each—living, working, studying, visiting.

The Union Gallery will be a *public building*. It will be open to the public and to University of Sydney students and alumni.

The Union Gallery will have a *commercial aspect* and must be self-supporting.

The Union Gallery will form a link between the University and the community, physically and conceptually, through both its location and its program.

Spaces	Description	Size
Galleries	There are two types of exhibitions to be shown: permanent and temporary	
	• Permanent collection of the University of Sydney Union	200 m^2 (50 m hanging space)
	• Temporary exhibitions	300 m^2 (75 linear meter of hanging space)

CRC Team Collaboration in High Bandwidth Virtual Environments
Face to face Page 2

	The total exhibition gallery area of 550 m^2 should include circulation space, wall thicknesses, services	Total: 500–550 m^2
Sculpture	• Sculpture space is also required for acquisitions by the permanent collection and for temporary exhibitions. This should be outdoor/indoor space	
Associated areas (with galleries)	Activities associated with the galleries	Total—approx. 450 m^2
	• Artwork store	200 m^2 (50 linear meter)
	• A workshop: for setting up exhibitions, curatorial work, repairs etc.	100 m^2
	• Exhibition catering: kitchen facilities adjacent to exhibition space for catering for exhibition openings	
	Services • Loading dock and possibly service court with good access to the galleries • One or two lifts are required for both service and daily passenger use. Maximum allowable is two lifts	
	• Offices for Director (large) for a curator administrative office and a receptionist • A boardroom with a large table for meetings of the ten members of the University of Sydney Students' Union Board meetings, pre-exhibition invited gatherings, etc, with adjacent small kitchen • Storage • Staff toilet, cleaners store [the size of a cubicle]	
Commercial/ merchandising areas	This is the commercial generator if the gallery and the interface between visitors, artists and the gallery. Foyer space must be provided, and security must be considered for after-hours activities	300–350 m^2
Outdoor areas	NOT included in the building footprint requirements	
Public toilets	Public toilets must be provided Visitors—on average 300 for an exhibition opening, and 1000 per exhibition, with twelve exhibitions annually	Toilets approx 40 m^2
Car parking	Not required	

B.3 Brief for Chap. 4 3D World Session

CRC Team Collaboration in High Bandwidth Virtual Environments
3D virtual world Page 1

Sydney College of Fine Arts Contemporary Dance School

Site—City Road—Darlington Road.

- THE PROGRAM

You are asked to design a contemporary building for a new Contemporary Dance School which is about to set up as part of University of Sydney College of Arts.

College of Arts has a series of key aims for the new Dance School

It will bring together the dance classes held in other locations

It will be the first school teaching contemporary dance in Sydney

it will be the only fine arts related school in Camperdown Campus

- YOUR CLIENT

Your client will be the University of Sydney College of Arts.

- THE SITE

The site is on the corner of City Road and Darlington Street, and is part of the Darlington campus. The site is currently occupied by the University Regiment and is defined on the accompanying plans.

Approval for any building on this site would be influenced by the University of Sydney Masterplan, and governed in the regulatory environment by the Local Environmental Plan and the Development Control Plan of South Sydney Council.

We have provided you with 2 × A3 of the site at 1:500, 1:200 and ortho-photo prints.

Approval and regulatory requirements

The site area is approximately 1800 m^2.

The maximum site coverage is 33 %. That is, 594 m^2, the building 'footprint', or built coverage of the site, excluding any external courts, sculpture display areas, unloading docks, service areas, etc. will be 500–550 m^2.

Building height limit is three storeys, and the building height and mass should enhance the urban context and respond appropriately to surrounding buildings and streetscape.

• THE BRIEF

The Dance School must have the following components:

Spaces	Area
Generic studio space 4 studios	200 m^2 each
Set store/workshop	280 m^2
Boardroom	18 m^2
Office space	75 m^2
Amenities—public/private	2 × 50 m
Foyer	250 m^2
Café	75 m^2
Sound control room	6 m^2
Courtyard—terrace—roof top space?	Undefined
Dressing room	70 m^2
Gym	70 m^2
Class room	40 m^2
Health spa—treatment room	25 m^2
Plant room—auditorium	20 m^2
Plant room—general	10 m^2

A car entry point is required—not so much as a loading dock. A pull-off point

B.4 Brief for Chap. 4 NetMeeting Session

Brief for a Harbourside Gallery	(Architect version)
During this design session you are asked to prepare a block model esquisse scheme for a proposed art/craft gallery on this site. The site and information below provide details about the location and use of the site. There are 30 min available for this investigation	
This project is to prepare a block model esquisse scheme for a proposed art/craft gallery on this site. You should assume that all existing buildings have been demolished before your scheme commences construction. There is no floor space ratio or height restriction applicable for this project so you may choose to liberate as much or as little of the site for open space as suits your scheme	

Site

The site is a triangular block as shown below. Site area is approximately 2800 m^2.

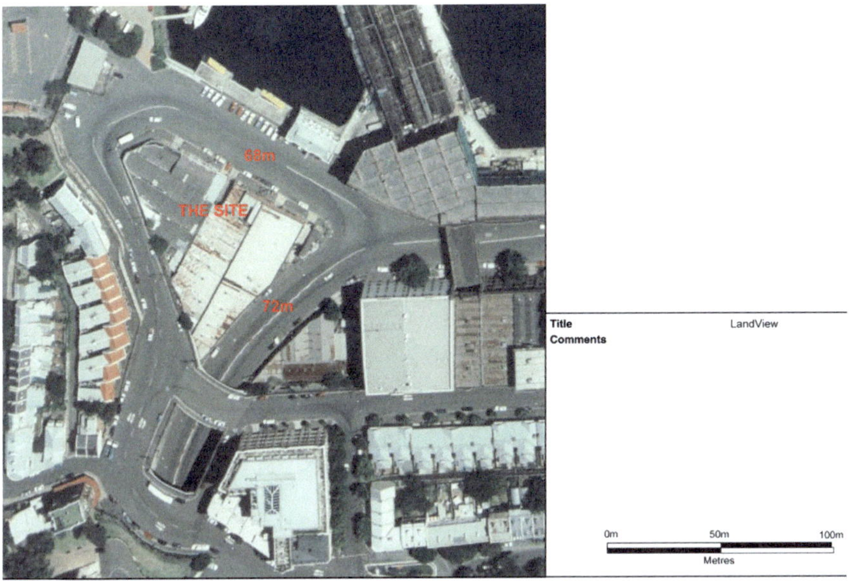

North is to the top of the picture where the harbour and wharves are visible. There are roads on all 3 sides of the site at varying heights relative to the ground floor of the existing buildings. You will see that one road crosses the other on an overhead bridge immediately to the south of the site, then ramps down along the west side of the triangle.

Accommodation required	Area (m^2)
Galleries and performance space	
Permanent exhibition suite	1500
Sculpture Garden	600
Temporary exhibition suite 1	750
Temporary exhibition suite 2	150
Forum	750
Front of house public areas	
Entrance/foyer	xx
Reception	30
Cloak store	20
Café (with after hours access)	200
Shop	100
Shop storage	30
Ticket office	25

(continued)

(continued)

Accommodation required	Area (m^2)
Members lounge	60
Back of house support areas	
Staff entry	xx
Loading dock	to suit truck 12.5 × 2.5 × 4.5 high
Unloading	60 (min. opening 4.5 × 4.5)
Bay for forklift	10
Exhibition receiving and preparation	200
Restoration and repair workshop	200
General storage	50
Chair storage	30
Cleaning	10
Board room	60
Director	30
Assistant directors and manager	20 × 3
Curators	15 × 12
Accounts	10 × 4
Security	20
Technical support	30 × 2
Volunteers	20
Toilets and showers	xx

Notes
No car parking required
Maximize energy efficiency and passive solar principles
All galleries to be naturally lit
Forum minimum span 25 metres
Separate delivery for café and shop

The participants were also given a collage of the photos showing the site and the surrounding area (Fig. B.1).

B.5 Brief for Chap. 6: House

The followings were excerpts from Bilda (2006).

Design Brief 01

Client: your task is to design a house for a couple, whose ages are 29 and 34. The female is a dancer, and the male is a painter. They are sensitive to colors and beauty, enjoying contact with the natural environment. In order to make their dream house come true, they have a budget of about $350,000.

Fig. B.1 Collage of site photos

Site: The site is located on the corner of the fully serviced home sites surrounded by a large central open-space recreation reserve in Matraville, one of Sydney's south eastern newly desirable locations. It is a trapezium in shape and slopes down to the edge of the recreation. The site has a view of the flame trees in the recreation reserve and the whole reserve. The site is 700 m^2. The floor space ratio for this site is 0.65:1, so the maximum floor plan can be 455 m^2.

House: the house is expected to be caressed by gentle sea breezes, and screened by a stately grove of magnificent flame trees along the edge of the estate. A sculpture garden is required for display of their art collections. According to the Randwick Development Control Plan No. 4, the height of a dwelling house should not exceed maximum of 9.5 m. Your task is to give forms to and arrange the following spaces on the site with the approximate sizes:

Living/dining area	40 m^2	Painter's studio	50 m^2
Kitchen	15 m^2	Dancer's studio	50 m^2
Bath	10 m^2	Observatory	20 m^2
Master bedroom	30 m^2	WC-shower	9 m^2
Bedroom	20 m^2	Parking space	36 m^2

Design Brief 02

Client: your task is to design a house for a re-married couple, whose ages are 42 (female) and 50 (male). The female is a part-time University lecturer, and the male is a Consultant and a Business Analyst. They've got 5 children (3 from previous marriages—ages 17, 15 and 13; 2 children of the current marriage—aged 7 and 5). They've got busy lifestyles and they also enjoy contact with the natural environment. The female works from home 2 days a week. The male invites colleagues from overseas every two months to their house for consulting purposes. There should be a study or work space, possibly shared by husband and wife. She will work from home, and he will need to use the space for meetings with colleagues. In order to make their functional, dream house come true, they have a budget of about $450,000.

Site: The site is located on the corner of the fully serviced home sites surrounded by a large central open-space recreation reserve in Matraville, one of Sydney's south eastern newly desirable locations. It is trapezium in shape and slopes down to the edge of the recreation. The site has a view of the flame trees in the recreation reserve and the whole reserve. The site is 700 m^2. The floor space ratio for this site is .65:1, so the maximum floor plan can be 455 m^2.

House: the house is expected to be caressed by gentle sea breezes, and screened by a stately grove of magnificent flame trees along the edge of the estate. A garden is required accommodating for children's recreational activities. According to the Randwick Development Control Plan No.4, the height of a welling house should not exceed maximum of 9.5 m. Your task is to give forms to and arrange the following spaces on the site with the approximate sizes (Fig. B.2):

Living/Dining area	40 m^2	Study/workspace	15 m^2
Kitchen	15 m^2	External play area	flexible
Bathroom	10 m^2	WC-shower	9 m^2
Master Bedroom	15–20 m^2	Parking space	36 m^2
Bedrooms arrangement for 5 children	70–120 m^2	Family room/children's accommodation	30 m^2

Design Discussion

The participants were interviewed after the blindfolded sessions, before they do a sketching session. They were asked open-ended questions which are listed below.

1. Can you describe how you went about the design process?
2. What role did talking play in the process?
3. How well developed do you think the design is?
4. If you were sketching in this session, do you think you would have produced a more developed, less developed or design of about the same level development?
5. How important is sketching in your design process?

Fig. B.2 Site layout and collage of the site photos

6. What role did visual or other imagery play in this design process?

Instructions

In this experiment we are interested in how you go about designing when you are allowed to and when you are not allowed to sketch while you design. To start with I will give you the brief for the design project and show you photographs illustrating the site and its surrounds. I also want you to memorize the brief. This will involve reading through the brief and then reciting it without looking at the document. You can do this a number of times until you feel comfortable that you can remember it. You can also ask to be reminded of specific aspects of the brief while you are designing.

Once you are satisfied about the brief, I will ask you to put on a blindfold and commence designing. However so that I can understand what you are doing I want you to TALK ALOUD when you are work on the design problem. What I mean by talk aloud that I want you to verbalize every thought. You do not describe what you are doing to me and don't judge whether it is important or not. Just keep saying what you are thinking while designing. It is not a conversation between you and me. Just act as if you are alone in the room speaking to yourself. It is OK if you stop talking for a short time. But if you are silent for a significant period of time, I will remind you to keep talking aloud. That is the only reason I am here. I am not here to judge what you say but to keep the experiment going properly.

Do you have any questions so far?

I have told you to think aloud, but we know that it is difficult and strange to think aloud. So we will start with a warm up exercise to practice. I want you to verbalize everything in your mind while you solve a simple problem. If you have any difficulties, please feel free to ask me. If I think there are some things I want, I will tell you. That is the reason for conducting the warm-up. Ok, let's start it.

The problem I want you to work on is to multiply two numbers in your head. That means, you cannot write anything down. It is not important whether the result is right or not. What I want is for you to practice verbalizing what you are thinking about as you solve the problem.

So verbalize everything in your mind while you multiply 24 times 34.

Good. Do you have any questions?

INSTRUCTIONS SPECIFIC TO BLINDFOLDED SESSION

OK. Now we are going to conduct the main experiment. In a moment I will give you a sheet with the design brief on it. Please pretend you are in a real situation, not hypothetical one. Once you begin to solve the design problem, please pretend I am not here. However, if you require some specific information just ask me.

The time limit will be 50 min. I will remind you when you have 5 min left. If you feel you have completed the design before this time let me know and we will stop the session then. Once you have completed the process, I want you to take off the blindfold and then as quickly as possible sketch out your design using any form of representation that you think is appropriate to document your design. Use sketches just to represent the design you have arrived at. Do not change it in any way. I will provide you with sheets of paper, the site plans, scale rule and so on that you will need to document your design.

To start the design process, you will be given the site plan and brief. Read through the brief aloud and then turn it face down and recite it back. Repeat this process until you feel you can remember the brief. Here is the brief and the site plan.

Now you have memorized the brief, look at the pictures of the site and its surrounds. When you have finished looking at these pictures you can if you want to re-read the brief to refresh your memory.

Ok. Now put on the blindfold and start designing.

Remember I will tell you when there is 5 min left to go and then when the design session is finished or you can tell me when you are happy with your design.

INSTRUCTIONS SPECIFIC TO SKETCHING SESSION

OK. Now we are going to conduct the main experiment. In a moment I will give you a sheet with the design brief on it. Please pretend you are in a real situation, not hypothetical one. Once you begin to solve the design problem, please pretend I am not here. However, if you require some specific information just ask me.

The time limit will be 50 min. I will remind you when you have 5 min left. If you feel you have completed the design before this time let me know and we will stop the session then.

To start the design process, you will be given the site plan and brief. Read through the brief aloud and then turn it face down and recite it back. Repeat this process until you feel you can remember the brief. Here is the brief and the site plan.

Now you have memorized the brief, look at the pictures of the site and its surrounds. When you have finished looking at these pictures you can if you want to re-read the brief to refresh your memory.

To start the design process, you will be given the site plan. In addition, you will be given sheets of tracing paper. Your design task should proceed with drawing a series of sketches. You can go onto a new sheet whenever you need to. In this way, you can try out different ideas, changing and developing the arrangement of spaces. By sketches we mean all kinds of drawings including just trifling traces of your pencil and small doodles besides main drawings. You are highly encouraged to draw anything. Begin by freely copying the outline of the site onto the first sheet of tracing paper, and numbering that page 1. Please number each sheet of tracing paper every time you go onto a new sheet. Try to remember why you drew things in each sketch and why you changed things in each subsequent sketch. Please save all your sketches (don't damage or erase), as we will ask you to describe them in the second part of the study. You can use as many sheets as you need to work out your design and you may overlay a new sheet on a previous sketch if you want to preserve some features of a previous sketch.

Appendix C
Experimental Setups

In the 3D world session (Chap. 4) and the NetMeeting session (Chap. 5), the designers' activities and verbal exchanges were recorded with a DVR system (digital video recording). In the face-to-face sessions (Chaps. 4 and 6) the activities were recorded by a digital video camcorder. Figure C.1 shows the set-up in the room with two participants using desktops in a collaborative design session. Two cameras and two desktop screens are input to the recording equipment.

Figure C.2 shows the set-up in the actual experiment setting. The placing of the cameras was an important issue, since we wanted to monitor all participant movement, verbalisation, gestures and the drawing actions and outcomes. Camera 1 and 2 captured the gestures, general actions such as walking, looking at, moving to the side etc. while the direct connections to the desktop screens captured the drawing process in detail. The DVR sys tem had a black box hard drive which includes 500 GB of storage space. Four views were inputed into the system. It was possible to see each camera view together on the monitor as the screen could be divided into four smaller views. Two cameras were used to monitor the two participants at their own locations. The other two inputs came from their desktop screens. There was one tie-clasp wireless microphone for one participant and an omni directional table top microphone for the other participant. The two microphone inputs were fed into the DVR system through a sound mixer. The sound was recorded as a backup during the sessions. That was done by connecting the mixer to the microphone input of a laptop. Figure C.3 shows the DVR system and the laptop.

NetMeeting includes a shared whiteboard application and web-cam application. The participants were able to see each other via the web-cam and also were able to talk to each other because they were located in the same room. Figure C.4 shows the physical configuration of the drawing surfaces. Before the experiment the participant were given a 15 min training session regarding the use of tangible

J.W.T. Kan and J.S. Gero, *Quantitative Methods for Studying Design Protocols*, DOI 10.1007/978-94-024-0984-0

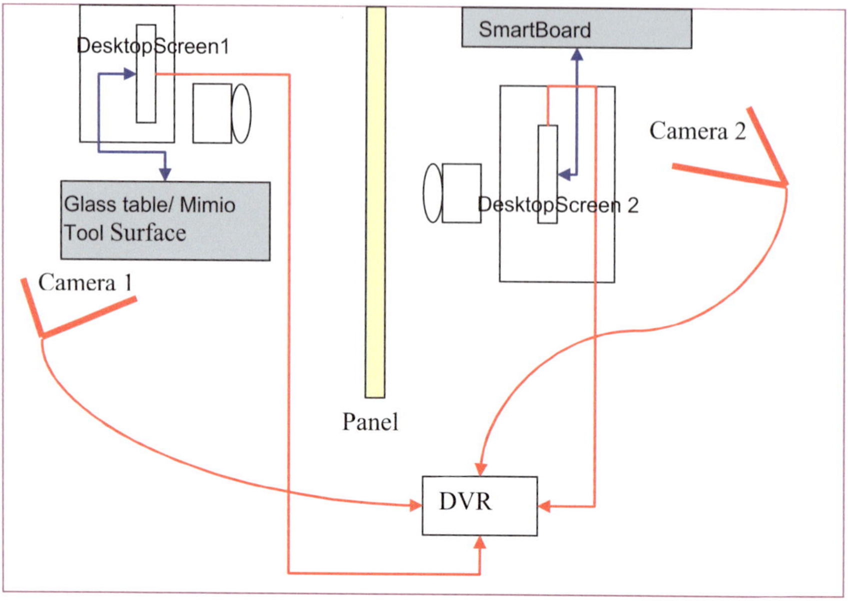

Fig. C.1 Diagram of set-up for NetMeeing and 3D world

Fig. C.2 *Left* Camera 1, desktop screen 1, and Mimio on glass table; *Right* Camera 2, desktop screen 2, and Smartboard

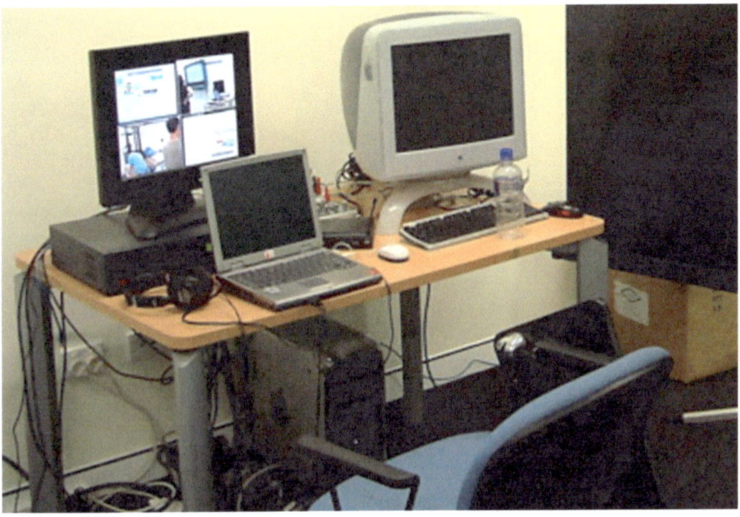

Fig. C.3 Video recording equipment

Fig. C.4 *Left* Mimio tool on glass table; *Right* Smartboard

interface of the Smartboard and Mimio tool. The design sessions were 30 min. In the session presented in the pilot study (Chap. 4), both participants used NetMeeting before.

In the 3D world session (Chap. 6) the participants used a customised 3D virtual world application—Active World. The 3D world included a multiuser 3D building environment, video contact, a shared whiteboard, and an object viewer/insert

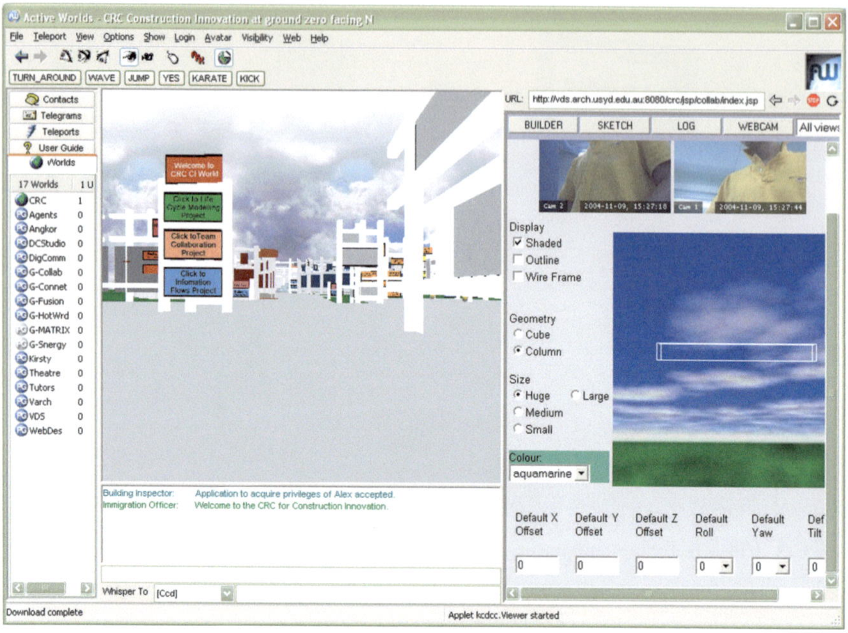

Fig. C.5 The Active World environment

feature. They were able to talk to each other because they were in the same room. A training session of 15 min was followed by a half an hour design session. Both participants in the analysis of Chap. 6 had some experience with this environment. Figure C.5 is a screen capture of the Active World.

Appendix D
Samples of Outcomes

D.1 Other Pages Produced During NetMeeting Session

See Fig. D.1.

D.2 Linkographs and Outcome of Chap. 6

See Figs. D.2, D.3, D.4, D.5, D.6, D.7, D.8, D.9, D.10, D.11, D.12 and D.13.

J.W.T. Kan and J.S. Gero, *Quantitative Methods for Studying Design Protocols*, DOI 10.1007/978-94-024-0984-0

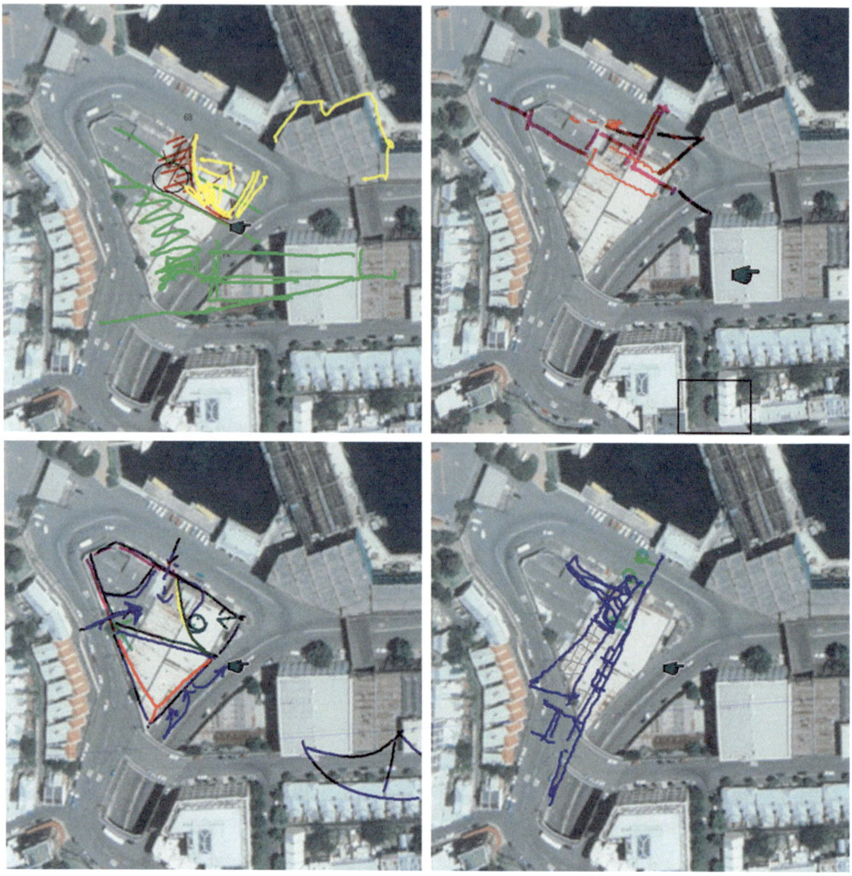

Fig. D.1 Other pages produced in the NetMeeting session (Chap. 5)

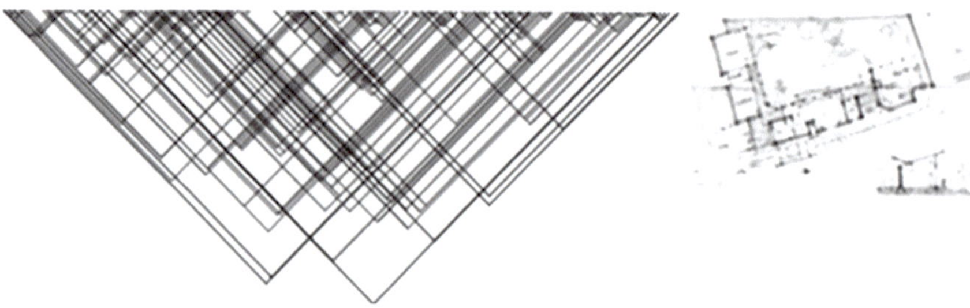

Fig. D.2 Linkographs and outcome of BF01

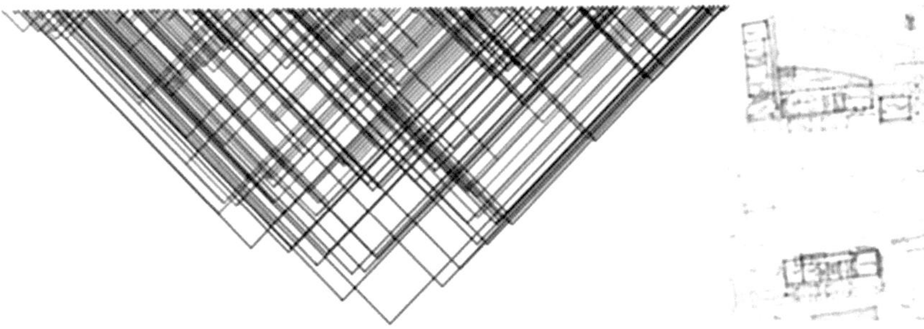

Fig. D.3 Linkographs and outcome of SK01

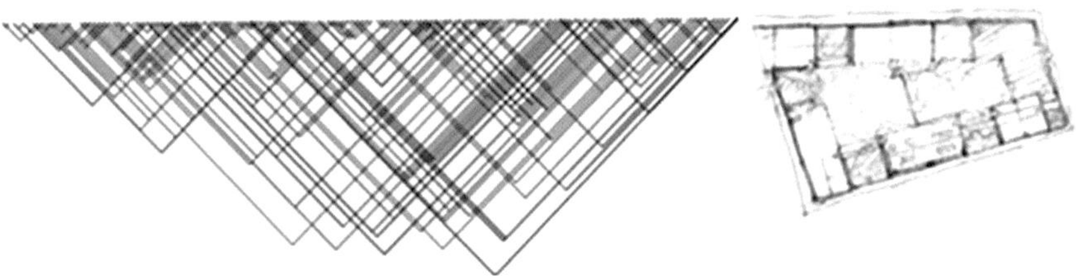

Fig. D.4 Linkographs and outcome of BF02

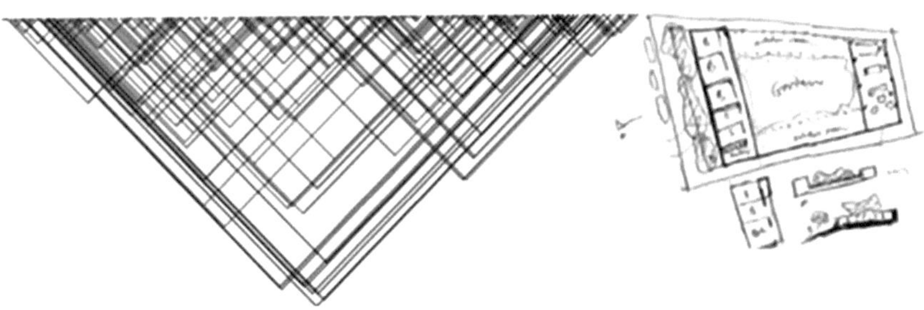

Fig. D.5 Linkographs and outcome of SK02

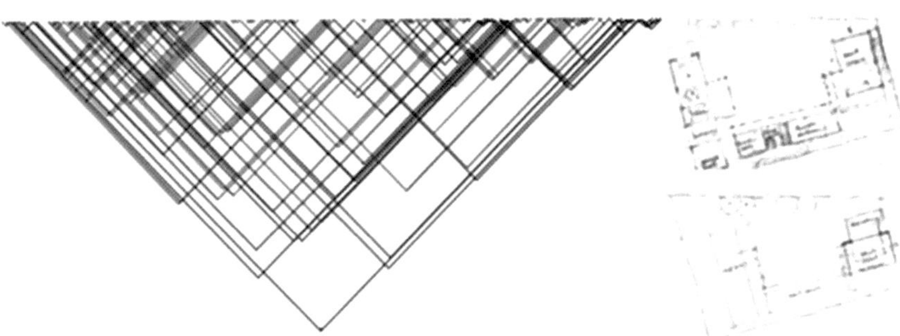

Fig. D.6 Linkographs and outcome of BF03

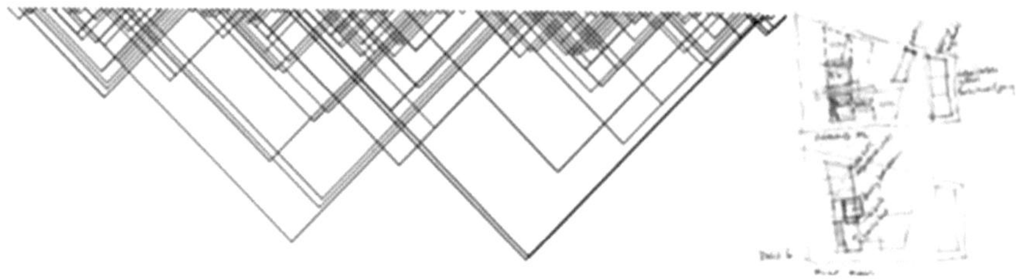

Fig. D.7 Linkographs and outcome of SK03

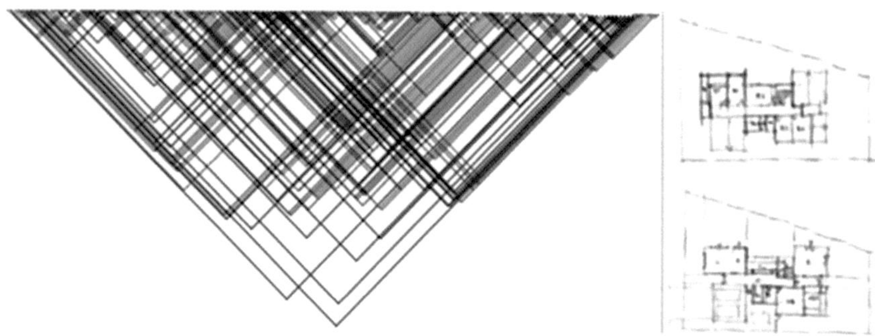

Fig. D.8 Linkographs and outcome of BF04

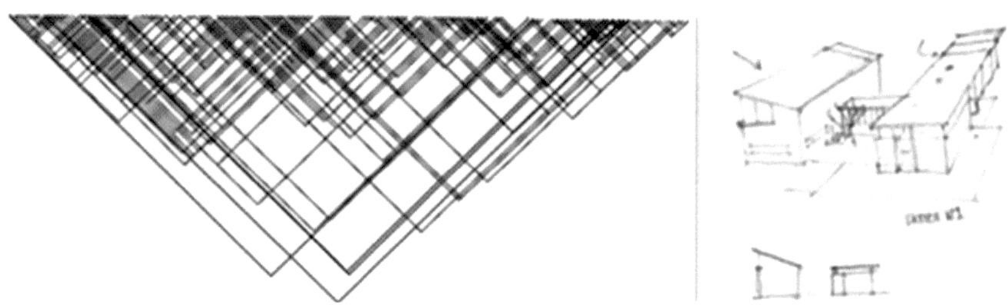

Fig. D.9 Linkographs and outcome of SK04

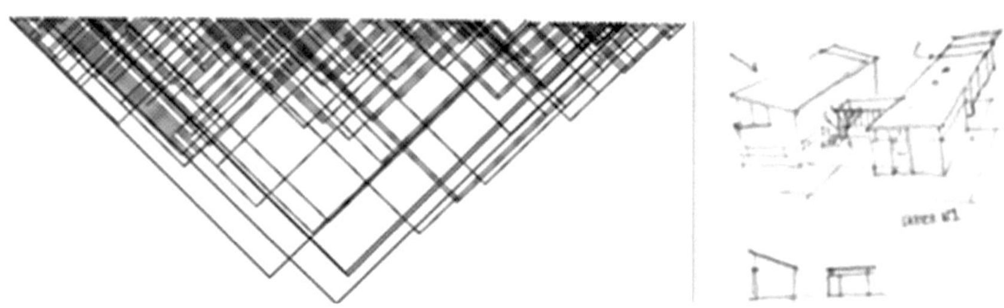

Fig. D.10 Linkographs and outcome of BF05

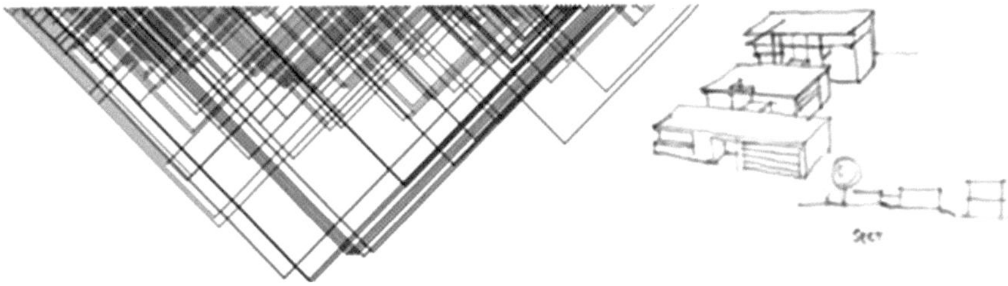

Fig. D.11 Linkographs and outcome of SK05

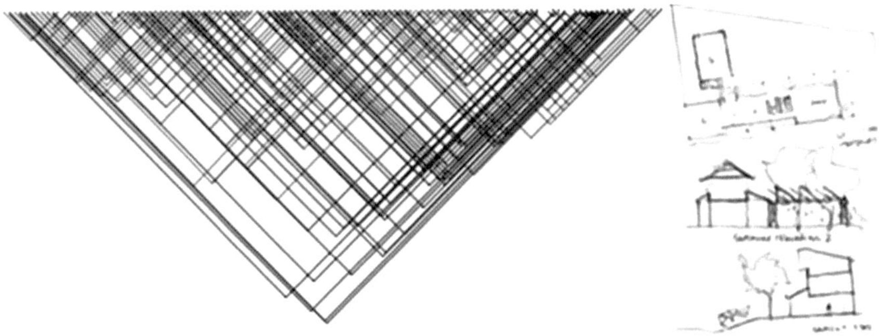

Fig. D.12 Linkographs and outcome of BF06

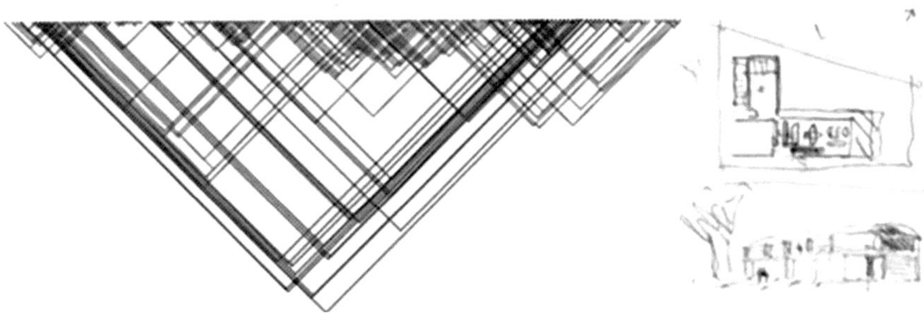

Fig. D.13 Linkographs and outcome of SK06

Appendix E
Situated FBS Processes

See Tables E.1, E.2 and E.3.

Table E.1 Derived processes of the brain storming session

Processes	Occurrence	Percentage (%)	Processes	Occurrence	Percentage (%)
$Be^i \rightsquigarrow Be^i$	69	6.61	$R \rightsquigarrow F^i$	1	0.10
$Be^i \rightsquigarrow B^i$	48	4.60	$R \rightsquigarrow R$	3	0.29
$Be^i \rightsquigarrow Fe^i$	1	0.10	$R \rightsquigarrow Si$	8	0.77
$Be^i \rightsquigarrow Se^i$	69	6.61	$Se^i \rightsquigarrow Be^i$	16	1.53
$Be^i \rightsquigarrow S^i$	5	0.48	$Se^i \rightsquigarrow B^i$	49	4.69
$Be^i \rightsquigarrow B^e$	16	1.53	$Se^i \rightsquigarrow B^e$	1	0.10
$Be^i \rightsquigarrow S^e$	2	0.19	Se^i Se^i	62	5.94
$B^i \rightsquigarrow Be^i$	54	5.17	$Se^i \rightsquigarrow S^i$	82	7.85
$B^i \rightsquigarrow B^i$	94	9.00	$Se^i \rightsquigarrow S^e$	45	4.31
$B^i \rightsquigarrow Fe^i$	3	0.29	$S^i \rightsquigarrow Be^i$	17	1.63
$B^i \rightsquigarrow F^i$	2	0.19	$S^i \rightsquigarrow B^i$	85	8.14
$B^i \rightsquigarrow Se^i$	9	0.86	$S^i \rightsquigarrow Se^i$	50	4.97
$B^i \rightsquigarrow S^i$	13	1.25	$S^i \rightsquigarrow S^i$	81	7.76
$B^i \rightsquigarrow B^e$	8	0.77	$S^i \rightsquigarrow R$	1	0.10
$Fe^i \rightsquigarrow Be^i$	12	1.15	$S^i \rightsquigarrow S^e$	14	1.34
$Fe^i \rightsquigarrow Fe^i$	4	0.38	$S^i \rightsquigarrow Fe^i$	1	0.10
$Fe^i \rightsquigarrow F^i$	6	0.57	$B^e \rightsquigarrow Be^i$	7	0.67
$Fe^i \rightsquigarrow S^i$	1	0.10	$B^e \rightsquigarrow B^e$	3	0.29
$F^i \rightsquigarrow Be^i$	1	0.10	$B^e \rightsquigarrow B^i$	2	0.19
$F^i \rightsquigarrow B^i$	2	0.19	$S^e \rightsquigarrow Be^i$	2	0.19
$F^i \rightsquigarrow Fe^i$	4	0.38	$S^e \rightsquigarrow B^i$	6	0.57
$F^i \rightsquigarrow R$	1	0.10	$S^e \rightsquigarrow Se^i$	14	1.34
$R \rightsquigarrow Be^i$	2	0.19	$S^e \rightsquigarrow S^i$	42	4.02
$R \rightsquigarrow Se^i$	2	0.19	$S^e \rightsquigarrow B^e$	2	0.19
$R \rightsquigarrow B^i$	13	1.25	$S^e \rightsquigarrow S^e$	9	0.86

J.W.T. Kan and J.S. Gero, *Quantitative Methods for Studying Design Protocols*, DOI 10.1007/978-94-024-0984-0

Table E.2 Derived processes of the face-to-face session

Processes	Occurrence	Percentage (%)	Processes	Occurrence	Percentage (%)
$B^i \rightsquigarrow B^i$	31	5.26	$R \rightsquigarrow F^i$	7	1.19
$B^i \rightsquigarrow Se^i$	14	2.38	$R \rightsquigarrow B^i$	2	0.34
$B^i \rightsquigarrow Be^i$	24	4.07	$R \rightsquigarrow S^i$	17	2.89
$Be^i \rightsquigarrow Be^i$	4	0.68	$R \rightsquigarrow Se^i$	4	0.68
$Be^i \rightsquigarrow S^i$	14	2.38	$R \rightsquigarrow Fe^i$	2	0.34
$Be^i \rightsquigarrow B^i$	14	2.38	$R \rightsquigarrow Be^i$	1	0.17
$Be^i \rightsquigarrow S^e$	9	1.53	$R \rightsquigarrow S^e$	4	0.68
$B^i \rightsquigarrow S^i$	10	1.70	$R \rightsquigarrow Be^i$	1	0.17
$Be^i \rightsquigarrow Se^i$	22	3.74	$R \rightsquigarrow R$	1	0.17
$Be^i \rightsquigarrow Be^i$	5	0.85	$S^i \rightsquigarrow R$	4	0.68
$B^i \rightsquigarrow R$	2	0.34	$S^i \rightsquigarrow S^i$	78	13.24
$B^i \rightsquigarrow F^i$	2	0.34	$S^i \rightsquigarrow Se^i$	22	3.74
$Be^i \rightsquigarrow F^i$	1	0.17	$Se^i \rightsquigarrow S^i$	17	2.89
$B^i \rightsquigarrow S^e$	7	1.19	$S^i \rightsquigarrow B^i$	30	5.09
$B^i \rightsquigarrow Fe^i$	1	0.17	$Se^i \rightsquigarrow Se^i$	13	2.21
$F^i \rightsquigarrow B^i$	1	0.17	$Se^i \rightsquigarrow B^i$	17	2.89
$F^i \rightsquigarrow F^i$	2	0.34	$S^i \rightsquigarrow Be^i$	11	1.87
$F^i \rightsquigarrow Fe^i$	2	0.34	$S^e \rightsquigarrow S^i$	46	7.81
$Fe^i \rightsquigarrow Be^i$	1	0.17	$S^e \rightsquigarrow B^i$	24	4.07
$F^i \rightsquigarrow Be^i$	1	0.17	$Se^i \rightsquigarrow Be^i$	3	0.51
$Fe^i \rightsquigarrow Se^i$	1	0.17	$Se^i \rightsquigarrow S^e$	14	2.38
			$S^i \rightsquigarrow S^e$	15	2.55
			$S^e \rightsquigarrow Se^i$	24	4.07
			$Se^i \rightsquigarrow Be^i$	2	0.34
			$S^e \rightsquigarrow S^e$	49	8.32
			$S^e \rightsquigarrow Be^i$	7	1.19
			$S^e \rightsquigarrow Be^i$	3	0.51
			$S^i \rightsquigarrow Be^i$	3	0.51

Table E.3 Derived processes of the 3D world session

Processes	Occurrence	Percentage (%)
$B^i \rightsquigarrow R$	1	1.32
$Be^i \rightsquigarrow Be^i$	1	1.32
$F^i \rightsquigarrow R$	1	1.32
$F^i \rightsquigarrow S^i$	1	1.32
$R \rightsquigarrow B^i$	4	5.26
$R \rightsquigarrow Sei$	3	3.95
$R \rightsquigarrow S^i$	3	3.95
$S^i \rightsquigarrow S^i$	33	43.42
$Se^i \rightsquigarrow Se^i$	17	22.37
$Se^i \rightsquigarrow B^i$	1	1.32
$Se^i \rightsquigarrow S^i$	7	9.21
$S^i \rightsquigarrow Se^i$	3	3.95
$Se^i \rightsquigarrow Be^i$	1	1.32

References

Akaike H (1973) Information theory as an extension of the maximum likelihood principle. In: Petrov BN, Csaki C (eds) Second international symposium on information theory. Akademiai Kiado, Budapest, pp 267–281

Akin O (1998) Cognition based computational design, The 30th anniversary celebration meeting of The Key Centre for Design Computing. University of Sydney, Australia

Amabile TM (1996) Creativity in context: update to "the social psychology of creativity". Westview Press, USA

Atman CJ, Bursic KM (1998) Documenting a process: the use of verbal protocol analysis to study engineering student design. J Eng Educ Spec Issue Assess 87(2):121–132

Atman CJ, Adams RS, Cardella ME, Turns J, Mosborg S, Saleem JJ (2007) Engineering design processes: a comparison of students and expert practitioners. J Eng Educ 96(4):359–379

Badke-Schaub P, Lauche K, Neumann A, Ahmed S (2007) Task team process: assessment and analysis of the development of shared representations in an engineering team. In: DTRS 7 design meeting protocols workshop proceedings, London, pp 97–109

Bakeman R, Gottman JM (1997) Observing interaction: an introduction to sequential analysis, 2nd edn. Cambridge University Press, New York

Bartlett FC (1932) Remembering: a study in experimental and social psychology. Cambridge University Press, Cambridge

Berlyne DE (1971) Aesthetics and psychobiology. Appleton-Century-Crofts, USA

Bilda Z (2006) The role of externalizations in design: designing in imagery versus sketching. PhD thesis, The University of Sydney

Bilda Z, Gero JS, Purcell T (2006) To sketch or not to sketch? That is the question. Des Stud 27(5): 587–613

Bly SA, Minneman SL (1990) Commune: a shared drawing surface. Office Information Systems, Cambridge, pp 184–192

Cai H, Do EYL, Zimring CM (2010) Extended linkography and distance graph in design evaluation: an empirical study of the dual effects of inspiration sources in creative design. Des Stud 31(2):146–168

Clancey WJ (1997) Situated cognition: on human knowledge and computer representations, learning in doing: social, cognitive, and computational perspectives. Cambridge University Press, Cambridge

Cross N (2004) Creative thinking by expert designers. J Des Res 4(2): online edition

Cross N (2007a) Designerly ways of knowing. Springer, London

Cross N (2007b) Foreward. In: McDonnell J, Lloyd P (eds) DTRS7 design meeting protocols: workshop proceedings, p ix

Cross N (2011) Design thinking: understanding how designers think and work. Berg, Oxford

J.W.T. Kan and J.S. Gero, *Quantitative Methods for Studying Design Protocols*, DOI 10.1007/978-94-024-0984-0

Cross N, Cross AC (1995) Observations of teamwork and social processes in design. Des Stud 16(2): 143–170

Cross N, Cross AC (1996) Observation of teamwork and social processes in design. In: Cross N, Christiaans H, Dorst K (eds) Analysing design activity. Wiley, Hoboken, pp 291–317

Cross N, Christiaans H, Dorst K (1996a) Introduction: the delft protocols workshop. In: Cross N, Christiaans H, Dorst K (eds), Analysing design activity. Wiley, Hoboken, pp 1–14

Cross N, Christiaans H, Dorst K (eds) (1996b) Analysing design activity. Wiley, Hoboken

Dewey J (1896) The reflex arc concept in psychology. Psychol Rev 3:357–370

Dong A (2009) The language of design: theory and computation. Springer, London

Dorst K (2003) The problem of design problems. In: Cross N, Edmonds E (eds) Expertise in design: design thinking research symposium 6. University of Technology, Sydney, pp 135–147

Dorst K (2004) On the problem of design problems-problem solving and design expertise. J Des Res 4(3):online edition

Dorst K, Dijkhuis J (1995) Comparing paradigms for describing design activity. Des Stud 16(2): 261–274

Dorst K, Dijkhuis J (1996) Comparing paradigms for describing design activity. In: Cross N, Christiaans H, Dorst K (eds) Analysing design activity, Wiley, Hoboken, pp 253–269

Dunlosky J, Metcalfe J (2009) Metacognition. SAGE Publications, New York

Eastman C (1970) On the analysis of intuitive design processes. In: GT Moore (ed) Emerging methods in environmental design and planning: proceedings of the design methods group first international conference, Cambridge, Mass, June 1968, pp 21–37

Eastman C (2001) New directions in design cognition: studies of representation and recall. In: McCracken M, Newsletter W, Eastman C (eds) Design knowing and learning: cognition in design education. Elsevier, pp 147–198

Ericsson KA, Simon HA (1993) Protocol analysis: verbal reports as data, revised edn. The MIT Press, Cambridge

Farbood M, Schoner B (2001) Analysis and synthesis of palestrina-style counterpoint using markov chains. In: Proceedings of international computer music conference, Havana, Cuba

Finke RA, Ward TB, Smith SM (1992) Creative cognition. MIT Press, Cambridge

Fischer T (2008) Obstructed magic: on the myths of observing designing and of sharing design observations. In: Teeraparbwong K, Nakapan W, Mahaek E, Nilkaew P (eds) CAADRIA2008: beyond computer-aided design, proceedings of the 13th conference on computer aided architectural design research in Asia

Flavell J (1979) Metacognition and cognitive monitoring: a new area of cognitive-development inquiry. Am Psychol 34(10):906–911

Gabriel GC (2000) Computer mediated collaborative design in architecture: The effects of communication channels in collaborative design communication (i.e. Communication). PhD thesis, University of Sydney

Galle P, Kovacs LB (1992) Introspective observations of sketch design. Des Stud 13(3):229–272

Gericke K, Schmidt-Kretschmer M, Blessing L (2007) Dealing with requirements: inferences on idea generation in the early stages of product development. In: McDonnell J, Lloyd P (eds), DTRS7 design meeting protocols: workshop proceedings, pp 127–140

Gero JS (1990) Design prototypes: a knowledge representation schema for design. AI Mag 11(4): 26–36

Gero JS (1999) Recent design science research: constructive memory in design thinking. Archit Sci Rev 42:3–5

Gero JS (2007) Ai edam at 20. Artif Intell Des Artif Intell Eng Des Anal Manuf 21:17–18

Gero JS (2010) Fixation while designing and its measurement. In: Badke-Schaub P, Cardoso C, Goncalves M (eds), Fixation or inspiration? Delft University of Technology, Delft, pp 1–9

Gero JS, Jiang H (2016) Exploring the design cognition of concept design reviews using the FBS-based protocol analysis. In: Adams RS, Siddiqui JA (eds) Analyzing design review conversations. Purdue University Press, West Lafayette. doi:10.5703/1288284315931

Gero JS, Kan JWT (2016) Empirical results from measuring design creativity: use of an augmented coding scheme in protocol analysis. In: Fourth international conference on design creativity. The Design Society (to appear)

Gero JS, Kannengiesser U (2000) Towards a situated function-behaviour-structure framework as the basis of a theory of designing. In: Smithers T (ed) Artificial intelligence in design'00, pp 1–5

Gero JS, Kannengiesser U (2002) The situated function-behaviour-structure. In: Gero JS (ed) Artificial intelligence in design'02, Kluwer Academic Publishers, pp 89–104

Gero JS, Kannengiesser U (2004) The situated function-behaviour-structure framework. Des Stud 25(4):373–391

Gero JS, Kannengiesser U (2014) The function-behaviour-structure ontology of design. In: Chakrabarti A, Blessing L (eds) An anthology of theories and models of design. Springer, Berlin, pp 263–283

Gero JS, McNeill T (1998) An approach to the analysis design protocols. Des Stud 19(1):21–61

Gero JS, Kannengiesser U, Pourmohamadi M (2014a) Commonalities across designing: Empirical results. In JS Gero (ed) Design computing and cognition'12, Springer, Berlin, pp 285–302

Gero JS, Kannengiesser U, Williams C (2014b) Does designing have a common cognitive behavior independent of domain and task: a meta-analysis of design protocols. In: International conference on human behavior in design

Goldschmidt G (1990) Linkography: assessing design productivity. In: Trappl R (ed) Cyberbetics and system'90, pp 291–298

Goldschmidt G (1992) Criteria for design evaluation: a process-oriented paradigm. In: Kalay YE (ed) Evaluating and predicting design performance, principles of computer-aided design. Wiley, Hoboken, pp 67–79

Goldschmidt G (1995) The designer as a team of one. Des Stud 16(2):189–209

Goldschmidt G (1996) The designer as a team of one. In: Cross N, Christiaans H, Dorst K (eds) Analysing design activity. Wiley, Hoboken, pp 65–91

Goldschmidt G (2014) Linkography: unfolding the design process. MIT Press, Cambridge

Goldschmidt G, Tatsa D (2005) How good are good ideas? correlates of design creativity. Des Stud 26(6):593–611

Gottman JM, Roy AK (1983) Sequential analysis: a guide for behavioral researchers. Cambridge University Press, Cambridge

Gunther J, Frankenberger E, Auer P (1996) Investigation of individual and team design processes. In: Cross N, Christiaans H, Dorst K (eds) Analysing design activity. Wiley, Hoboken, pp 117–131

Hay L, McTeague C, Duffy A, Pidgeon L, Vuletic T, Grealy M (2016) A systematic review of protocol studies on conceptual design cognition. In: Gero JS (ed) Design computing and cognition'16, Springer, Berlin, pp 143–162

Hughes J, Parkes S (2003) Tends in the use of verbal protocol analysis in software engineering research. Behav Inf Technol 22(2):127–140

Ingle BR (2013) Design thinking for entrepreneurs and small businesses: putting the power of design to work. Apress, Berkeley

Jiang H (2012) Understanding senior design students' product conceptual design activities. PhD Thesis, National University of Singapore, Singapore

Jiang H, Gero JS, Yen CC (2014) Exploring designing styles using Problem-Solution indexes. In: Gero JS (ed) Design computing and cognition'12, Springer, Berlin, pp 85–101

Jones CJ (1970) Design methods: seeds of human futures. Wiley, London

Kan JWT, Gero JS (2004) A method to analyse team design activities. In: Bromberek Z (ed) Contexts of architecture: proceedings of the 38th annual conference of the Architectural Science Association ANZAScA and the International Building Performance Simulation Association, School of Architecture, University of Tasmania, Tasmania, Australia, pp 111–117

Kan JWT, Gero JS (2008a) Acquiring information from linkography in protocol studies of designing. Des Stud 29(4):315–337

Kan JWT, Gero JS (2008b) Do computer mediated tools affect team design creativity? In: Nakapan W, Mahaek E, Teeraparbwong K, Nilkaew P (eds) CAADRIA 2008. Pimniyom Press, Chiang Mai, pp 263–270

Kannengiesser U, Gero JS, Wells J, Lammi M (2015) Do high school students benefit from pre-engineering design education? In: International conference on engineering design—ICED2015

Kavakli M, Gero JS (2002) The structure of concurrent cognitive actions: a case study of novice and expert designers. Des Stud 23(1):25–40

Kavakli M, Gero JS (2003) Strategic knowledge differences between an expert and a novice designer: an experimental study. In: Lindemann U (ed) Human behaviour in design. Springer, Berlin, pp 42–51

Keinonen T (2006) Introduction to concept design. In: Keinonen T, Takala R (eds) Product concept design: a review of the conceptual design of products in industry. Springer, Berlin, pp 1–31

Kemeny JG, Snell JL (1960) Finite Markov chains. The University Series in Undergraduate Mathematics, D. Van Nostrand Company, Inc., Princeton

Kenner H, O'Rourke J (1984) A travesty generator for micros: nonsense imitation can be disconcertingly recognizable. BYTE 12:449–469

Krippendorff K (1986) Information theory: structural models for qualitative data, quantitative applications in social sciences. Sage, Beverly Hills

Kruger C, Cross N (2006) Solution driven versus problem driven design: strategies and outcomes. Des Stud 27(5):527–548

Kvan T, Yunyan J (2004) Students' learning styles and their correlation with performance in architectural design studio. Des Stud 26(1):19–34

Langville AN, Meyer CD (2006) Updating markov chains with an eye on Google's PageRank. SIAM Journal on Matrix Analysis and Applications 27(4):968–987

Lawson B (1997) Design in mind. Architectural Press, Oxford

Lawson B, Loke SM (1997) Computers, words and picture. Des Stud 18(2):171–183

LINKODER (2011) http://www.linkoder.com/

Lloyd P (2007) Ethical imagination and design. In: McDonnell J, Lloyd P (eds) DTRS7 design meeting protocols: workshop proceedings, pp 45–56

Lockwood T (ed) (2009) Design thinking: integrating innovation, customer experience and brand value. Allworth Press, New York

Logie RH (1995) Visuo-spatial working memory. Lawrence Erlbaum Associates, Publishers, USA

Logie RH (2001) Working memory: a mental space for design and discovery. In: Gero JS, Tversky B, Purcell T (eds) Visual and spatial reasoning in design II, University of Sydney, Sydney, pp 223–235

Lopez-Mesa B, Thompson G (2006) On the significance of cognitive style and the selection of appropriate design methods. J Eng Des 17(4):371–386

MacKay DJC (2003) Information theory, inference and learning algorithms. Cambridge University Press, Cambridge

Maher ML, Bilda Z, Gul LF (2006) Impact of collaborative virtual environments on design behaviour. In: Gero J (ed) Design computing and cognition'06. Springer, Dordrecht, pp 305–321

Mazijoglou, M, Scrivener, S and Clark, S: 1996, Representing design workspace activity, in N Cross, H Christiaans and K Dorst (eds), Analysing Design Activity, Wiley, Hoboken, pp 389–416

McDonnell J, Lloyd P (eds) (2007) DTRS7 design meeting protocols: workshop proceedings
McNeill T, Gero JS, Warren J (1998) Understanding conceptual electronic design using protocol analysis. Res Eng Des 10(3):129–140
Miller GA (1956) The magical number seven, plus or minus two: Some limits on our capacity for processing information. Psychol Rev 63:81–97
Minneman SL (1991) The social construction of a technical reality: empirical studies of the social activity of engineering design practice. Master's thesis, Stanford University, Stanford
Murty P (2006) Discovery processes in designing. PhD thesis, Key Centre of Design Computing and Cognition, Faculty of Architecture, Design and Planning, University of Sydney
Oak A (2011) What can talk tell us about design? analyzing conversation to understand practice. Des Stud 32(3):211–234
Olson GM, Olson JS (2000) Distance matters. Hum-Comput Interact 15(2/3):130–178
Paivio A (1986) Mental representation: a dual coding approach. Oxford University Press, Oxford
Perry GT, Krippendorff K (2013) On the reliability of identifying design moves in protocol analysis. Des Stud 34(5):612–635
Plattner H, Meinel C, Leifer L (2012) Design thinking research: studying co-creation in practice. Springer, New York
Purcell AT, Williams P, Gero JS, Colbron B (1993) Fixation effects: do they exist in design problem solving? Environ Plann B 20:333–345
Riegler A (2005) Constructive memory. Kybernetes 34(1/2):89–104
Sarkar P, Chakrabarti A (2011) Assessing design creativity. Des Stud 32(4):348–383
Schacter DL, Addis DR (2007) The cognitive neuroscience of constructive memory: remembering the past and imagining the future. Philos Trans R Soc B 362:773–786
Schon DA (1991) The reflective practitioner: how professionals think in action. Ashgate, Avebury
Schon DA, Wiggins G (1992) Kinds of seeing and their functions in designing. Des Stud 13(2):135–156
Shannon CE (1948) A mathematical theory of communication. Bell Syst Tech J 27:397–423
Simon HA (1969) The science of artificial. MIT Press, Cambridge
Song T, Becker K, Gero J, Deberard S, Lawanto O, Reeve E (2016) Problem decomposition and recomposition in engineering design: a comparison of design behavior between professional engineers, engineering seniors, and engineering freshmen. J Technol Educ 27(2):37–56
SPSS: SPSS for windows release 11.5.0
Stempfle J, Badke-Schaub P (2002) Thinking in design teams an analysis of team communication. Des Stud 23(5):473–496
Suwa M, Purcell T, Gero JS (1998) Macroscopic analysis of design processes based on a scheme for coding designers' cognitive actions. Des Stud 19(4):455–483
Suwa M, Purcell T, Gero JS (2000) Unexpected discoveries and s-invention of design requirements: important vehicles for a design process. Des Stud 21(6):539–567
Tang HH (2002) Exploring the roles of sketches and knowledge in design process. PhD thesis, Key Centre of Design Computing and Cognition, Faculty of Architecture, University of Sydney
Tang HH, Gero JS (2002) A cognitive method to measure potential creativity in designing. In: Bento C, Cardoso A, Wiggins G (eds) Workshop 17 creative systems: approaches to creativity in AI and cognitive science, ECAI-02, pp 47–54
Toker F (2003) Fallingwater rising: Frank Lloyd Wright, E. J. Kaufmann, and America's most extraordinary house. Knopf, New York
Valkenburg R, Dorst K (1998) The reflective practice of design teams. Des Stud 19(3):249–271
van der Lugt R (2000) Developing a graphic tool for creative problem solving in design groups. Des Stud 21(5):505–522
van der Lugt R (2003) Relating the quality of the idea generation process to the quality of the resulting design ideas. In: International conference on engineering design (ICED). Stockholm, Sweden, pp 19–21

van Someren MW, Barnard YF, Sandberg JA (1994) The think aloud method: a practical guide to modelling cognitive processes, knowledge based systems. Academic Press, USA

von Mayrhauser A, Lang S (1999) A coding scheme to support systematic analysis of software comprehension. IEEE Trans Softw Eng 25(4):526–540

Yilmaz S, Daly S, Seifert S, Gonzalez R (2015) How do designers generate new ideas? Design heuristics across two disciplines. Des Sci 1:e4. doi:10.1017/dsj.2015.4

Yu R, Gero JS, Gu N (2013) Impact of using rule algorithms on designers' behavior in a parametric design environment: preliminary results from a pilot study. In: Zhang J, Sun C (eds) Global design and local materialization—CAADFutures, Springer, Berlin, pp 13–22

Yu R, Gu N, Ostwald M, Gero JS (2015) Empirical support for problem-solution co-evolution in a parametric design environment. AIEDAM 25(1):33–44

Yukhina E (2007) Cognitive abilities & learning styles in design processes and judgements of architecture students. PhD thesis, Faculty of Architecture, The University of Sydney

Zolin R, Hinds PJ, Fruchter R, Levitt RE (2004) Interpersonal trust in cross-functional, geographically distributed work: a longitudinal study. Information and Organization 14(1):1–26

MIX
Papier aus verantwortungsvollen Quellen
Paper from responsible sources
FSC® C105338

If you have any concerns about our products,
you can contact us on
ProductSafety@springernature.com

In case Publisher is established outside the EU,
the EU authorized representative is:
Springer Nature Customer Service Center GmbH
Europaplatz 3, 69115 Heidelberg, Germany

Printed by Libri Plureos GmbH
in Hamburg, Germany